D0983644

JOHN CLARE AND
PICTURESQUE LANDSCAPE

JOHN CLARE
AND
PICTURESQUE
LANDSCAPE

TIMOTHY BROWNLOW

CLARENDON PRESS · OXFORD

1983

Oxford University Press, Walton Street, Oxford OX2 6DP
London Glasgow New York Toronto
Delhi Bombay Calcutta Madras Karachi
Kuala Lumpur Singapore Hong Kong Tokyo
Nairobi Dar es Salaam Cape Town
Melbourne Auckland
and associates in
Beirut Berlin Ibadan Mexico City Nicosia

Oxford is a trade mark of Oxford University Press

Published in the United States by
Oxford University Press, New York

British Library Cataloguing in Publication Data

Brownlow, Timothy
John Clare and picturesque landscape.
1. Clare, John 1793-1864—Criticism and
interpretation 2. Landscape in literature
I. Title
821'.7 PR4453.C6
ISBN 0-19-812808-8

Typeset by DMB (Typesetting), Oxford,
and printed in Great Britain
at the University Press, Oxford
by Eric Buckley
Printer to the University

IN
LOVING MEMORY
OF
MY FATHER
AND
MY MOTHER
who always had faith

ACKNOWLEDGEMENTS

My thanks are due, first, to Professor Eric Robinson, who holds the copyright of Clare's unpublished works. With his permission, I have been able to quote from the typescript of the forthcoming Clare volume in the Oxford Poets series, and from the transcripts which will form the basis of the Oxford English Texts edition. I am also grateful for Professor Robinson's general encouragement. For expert and cheerful help in threading my way through the labyrinths of Clare manuscripts, I am greatly indebted to David Powell. I am indebted to the staff of the Peterborough Museum; George Atkinson, Miss Arnold, and the staff of the Northampton Public Library (this library deserves special mention for its unique and constantly growing Clare collection), the staffs of the Bodleian Library, Dalhousie University Library, Harvard College Library, Saint Mary's University Library, and the Library of York University, Toronto. I also record my debt to Mrs Beveridge and the Library of Acadia University for allowing me extended loans of indispensable Clare volumes in their possession. I am also indebted to the staff of the Usher Art Gallery, Lincoln for providing information about Peter De Wint, and to Francis Greenacre, from whom I received valuable help connected with E. V. Rippingille and Francis Danby.

Norman Jeffares helped me to find my direction, Desmond Maxwell was an astute and hospitable supervisor in the early stages of my work, John Unrau taught me much and encouraged me to make some crucial decisions, William Whitla and Brian Hepworth were rigorous and inspiring teachers, and I have learned much from informal discussions with Maurice Elliott, Michael Collie, and Theodore Heinrich. W. J. Keith was an informed and gentle external examiner when I presented an earlier version of this book as a thesis at York University, Toronto. His comments were a substantial help to me as I proceeded to convert the thesis into a book.

Bruce Stovel read the typescript with a keen eye for detail; his comments did much to improve my text. I am also indebted to Alan Kennedy, Francis Fox, and Ann Stone, who read the work at various stages and who gave me the benefit of their opinions. James Gray was always supportive. Shelagh Kennedy did useful research work on my behalf at the British Museum and the Royal Academy. For excellent typing services I must thank Pat Lucas, Sally Owen, Bette Tetreault, Chitra Paranjape, Judith Fox, and Deborah Maskell.

I am indebted to the Marquess of Exeter and the Earl Fitz-william for permission to explore the gardens and parks of Burghley House and Milton House respectively. Alec Whitby and Daniel Crowson were my charming hosts on these occasions. The late G. E. L. Owen was a delightful host at King's College, Cambridge, where I was able to discuss Clare matters with John Barrell.

Jon Stallworthy has repeatedly refreshed my flagging spirits with friendship and astute advice, Francis Warner has been warmly hospitable and encouraging, Peter Levi has shared some of his wide knowledge and his deep love of Clare, Bernard Richards has provided enthusiasm, advice, and hospitality, and David Fleeman has been a model of punctilious scholarship and generous encouragement. I would also like to thank J. R. Watson for his friendly correspondence and Jay Appleton for inspiration and advice. Dick and Tam Frost lightened my load with their delightful and generous friendship. I am also indebted in various ways to John and Chris Kelly, Michael Herren, Douglas Gray, Ron Tetreault, Bernard Hickey, Michael Barker, John Press, Kraft von Maltzahn, David and Elizabeth Jones, Glynne Wickham, Michael Wheeler Booth, Christopher and Adèle Crowder, Malcolm Ross and John Ettlinger.

David FitzGerald, my former colleague at Saint Columba's College, Rathfarnham, Co. Dublin, first made me aware of Clare's uniqueness and he lent me the book which was to change the course of my life. Paddy Somerville-Large has been a charming and generous host at Vallombrosa, Co. Wicklow, a house which was a second home to me for many years. To Admiral and Mrs W. M. Landymore I am indebted for their continual interest and generosity. My wife, Jennifer, has lived for over eight years now with John Clare as a house guest; she has extended to his ghostly presence the same charity and loving attention which she bestows on me.

That Edward Malins will not read this book is a grievous loss. To him, and to his wife Meta, I owe the memory of a short but precious friendship. I am the better for having shared his urbanity, wide knowledge, passion for the arts, and not least, mischievous sense of humour. 'Farewell, too little and too lately known'.

I must also thank York University, Toronto, for its award of a Graduate Assistantship, Stong College, York University, Toronto—especially the former Master, Virginia Rock—for providing me with a home for two years, the government of Ontario for the award of a Queen Elizabeth II Ontario Scholarship, the Canada Council for providing a Doctoral Fellowship, Pembroke College, Oxford, for its hospitality to a Migrant Graduate for two years, the Izaak Walton Killam Trust at Dalhousie University for the award of a Post-Doctoral Fellowship, the Research Development Fund of Dalhousie University for two grants in aid of research, Saint Mary's University for the award of a Travel Grant, and my colleagues at Saint Mary's University, in particular Tom Musial, for their encouragement.

Part of Chapter One appeared in a slightly different version, in *University of Toronto Quarterly*, Fall 1978.

Thomas Gray's drawings of moths are reproduced by permission of the Houghton Library, Harvard University.

John Clare's drawings of shells and a moth come from the Northampton MS 15.

The author and publisher wish to thank the following for permission to reprint extracts from unpublished versions of Clare's poetry: Eric Robinson and the Peterborough Museum Society for the extracts from manuscript sources in the Peterborough Museum (from 'Helpstone', PMS, A3-25; from a poem on Burghley Park, PMS, A6-38; from the Pickworth Elegy, PMS, A3-87; from 'Hollywell', PMS, B2-127/127a, 128; from 'The Woodman', PMS, B2-238a/239; 'The Mole', PMS, A54-400;

prose extracts, PMS, A58-10, PMS, A46-21; 'Mouse's Nest', PMS, A61-6; from 'Walcott Hill and Surrounding Scenery', PMS, B7-38c, B7-40b, 41b, 42b, 43b; 'November', PMS, A61-47; 'Deluge', PMS, A61-79; 'Fir-wood', PMS, A57-1; 'The Tulip and the Bee', PMS, A51-92); Eric Robinson, George Atkinson, and Northamptonshire Libraries for the extracts from manuscript sources in the Northampton Public Library (from a poem on Burghley Park, NMS, 1-22; from 'The Village Minstrel', NMS, 3-167, 3-169, 3-170; from 'Summer Evening', NMS, 1-1; 'To the Ivy', NMS, 32-29; prose extracts, NMS, 7, NMS, A42-131); and Eric Robinson and the Trustees of the British Library for the extract from the Rippingille MS (BL MS EG. 2246, fo. 366).

Thanks are also due to A. J. V. Chapple and Yale University for the manuscript fragment quoted from *The Yale University Library Gazette*, 31.1 (July 1956), p. 48; Geoffrey Grigson for the manuscript fragment quoted from *The Mint* (1946), p. 175; Anne Tibble and R. K. R. Thornton and the Mid Northumberland Arts Group for extracts from *The Midsummer Cushion* (1979) ('Evening Primrose', *MC*, p. 432; 'To Dewint', *MC*, p. 404; from 'Pleasant Spots', *MC*, p. 452; from 'The Yellowhammers Nest', *MC*, p. 239; from 'The Yellow Wagtails Nest', *MC*, p. 212; from 'Pastoral Poesy', *MC*, p. 292).

In addition the following bodies are thanked for their permission to quote from published sources: *Selected Poems and Prose of John Clare*, ed. Eric Robinson and Geoffrey Summerfield, © Eric Robinson 1967. By permission of Curtis Brown Academic Ltd. *The Shepherd's Calendar* by John Clare, ed. Eric Robinson and Geoffrey Summerfield, © OUP 1964. By permission of Oxford University Press. Glossary © Eric Robinson, 1983.

CONTENTS

LIST OF ABBREVIATIONS

BL MS	British Library manuscript
Cowper: PW	*Cowper: Poetical Works*, ed. H. S. Milford
Essay	*Essay on the Application of Natural History to Poetry*, John Aikin
Essays	*Essays on the Picturesque*, Uvedale Price
Francis Danby	*Francis Danby: Varieties of Poetic Landscape*, Eric Adams
Heritage	*Clare: The Critical Heritage*, ed. Mark Storey
The Idea of Landscape	*The Idea of Landscape and the Sense of Place 1730-1840: an Approach to the Poetry of John Clare*, John Barrell
Letters	*The Letters of John Clare*, ed. J. W. and Anne Tibble
MC	*The Midsummer Cushion*, ed. Anne Tibble and R. K. R. Thornton
NMS	Northampton manuscript
PMS	Peterborough manuscript
Poems	*The Poems of John Clare*, ed. J. W. Tibble
Pope: Poems	*The Poems of Alexander Pope*, ed. John Butt
Prose	*The Prose of John Clare*, ed. J. W. and Anne Tibble
SC	*John Clare: The Shepherd's Calendar*, ed. Eric Robinson and Geoffrey Summerfield
Selborne	*The Natural History of Selborne*, Gilbert White
SPP	*Selected Poems and Prose of John Clare*, ed. Eric Robinson and Geoffrey Summerfield
Thomson: PW	*James Thomson: Poetical Works*, ed. J. Logie Robertson
Wordsworth: PW	*The Poetical Works of William Wordsworth*, ed. Ernest de Selincourt

An asterisk in the quotations indicates that the word is explained in the Glossary on p. 153.

Numbers given after the abbreviated source reference at the end of quotations denote the page reference, preceded by volume number where applicable.

Introduction

Much attention has been paid in recent years to the poetry Clare wrote in the Northampton Asylum between 1841 and 1864. This poetry has an intensity and visionary quality different from but not necessarily superior to the early descriptive work, whose virtues are often overlooked by modern critics. This book is an attempt to outline those virtues by examining Clare's verse and prose from the first signs of his maturity (c.1821) to the beginning of the Asylum period (1841). My concern is with Clare's use of landscape and natural history, not with his prolific output of Ballads and Songs. I take as my motto Ian Jack's words, 'before Clare was a mad poet, he was a sane poet',[1] and my book sets out to draw attention to the sanity and delight of Clare's early work.

I have tried to steer a course between what I see as two critical pitfalls: treating Clare either as a Peasant Poet or as a failed Blake. The former patronizing attitude was the habit of many critics until recently. The latter danger is more in evidence in the approach to Clare typified by Harold Bloom. His attempt to ascribe great Romantic qualities to Clare has the effect of proving how much greater Blake's and Wordsworth's Romantic qualities are. In fact, he admits Clare is a 'Wordsworthian shadow'.[2] The emphasis here is unfortunate. Bloom makes some amends in *The Oxford Anthology of English Literature* (1973). Although the editors (of whom Bloom is one) still compare Clare with Wordsworth and Blake, they admit that he 'is all but a major poet in his own right', and that in his best poems, 'he matches the major Romantics' (*Romantic Poetry and Prose*, pp. 559-60). Horace Gregory's judgement is more balanced:

In learning John Clare's language some patience is required. Judge Clare by a Wordsworthian scale of values, and it would seem that he had no philosophy; compare him with Byron and he would be found lacking in brilliance and personality; contrast him with Keats and he seems provincial ... In other words, Clare founded his esthetic upon a premise that refuted the work of his contemporaries.[3]

Some critics assume that Clare had no very clear ideas or aesthetic aims, but the more one studies the poems, the prose, and the manuscript notebooks the less tenable this assumption becomes. Clare's woodnotes may have been native, but they were very far from wild. It is often thought that Clare's grammatical aberrations are the result of lack of education or sloppiness, but he held strong convictions about the form his poems should take. The evidence of his letters is that he resented John Taylor's alterations of his manuscripts, though not in a position to have his own way. He writes to Taylor in 1822, '...grammer in learning is like tyranny in government—confound the bitch I'll never be her slave & have a vast good mind not to alter the verse in question ...'.[4] Clare had similar views about the 'improvement' of the landscape. It is my belief that to tidy up Clare's literary 'landscape' is to lose contact with that ancient open landscape which he loved; we should encounter both unadorned, unenclosed, and unimproved. This relationship between syntax and vision was stressed by Capability Brown, who described his art to Hannah More in literary terms and compared his landscape alterations to punctuation.[5]

In asking what Clare could do, I have avoided words such as great, major, and minor. These judgements can only be made when we have a more representative and accurate text. Auden puts Clare in with the minor poets and writes:

To qualify as major, a poet, it seems to me, must satisfy about three and a half of the following five conditions: 1). He must write a lot 2). His poems must show a wide range in subject matter and treatment 3). He must exhibit an unmistakable originality of vision and style 4). He must be a master of verse technique 5). In the case of all poets, we distinguish between their juvenilia and their mature work, but, in the case of the major poet, the process of maturing continues until he dies ...[6]

I would say that Clare earns a full mark for Nos. 1). and 3)., and a half-mark for Nos. 2)., 4)., and 5). respectively. That brings the total to three and a half, but I leave it to others to juggle with categories and definitions. One thing is certain; Clare's rating stands higher today than ever before.

In a lecture given at the University of London (16 January 1975), Peter Levi struck an exemplary balance:

Nine tenths of our interest in literature is because we are hungry for something genuine ... What was important in John Clare's genuineness was neither the extremity of his madness nor the sweetness and harshness of his rural youth. They do mark him and limit him and define him. But in his artistry his workshop was the English language, and what is genuine in him can be seen and felt as language. That is the only medium in which we know him.

What could Clare do with this medium, oral and literary, inherited from a long tradition, but more directly from the eighteenth century, from Pope, Thomson, Gray, Cowper, Crabbe, Goldsmith, and Wordsworth? (All these authors were in Clare's library.)[7] The following chapters attempt to answer this question by examining Clare's relationship to prospect poetry, his knowledge of natural history, his solution of the problems of descriptive poetry, his knowledge of painting, and his radical definition of the word 'taste'. My aim is to show how Clare absorbed and modified the eighteenth-century tradition in a unique way, so that he is misjudged if he is seen either as the culmination of a worn-out mode or as a lesser star in the Romantic galaxy.

Clare does not fit easily into any historical category. To appreciate his uniqueness, we have to understand the tradition of Landscape, and that involves us in the Picturesque and the Topographical. And if Pastoral relies for its effect upon the juxtapositions of urbanity and simplicity, formal language and innocent sentiments, a leaden present and a golden past, then Clare is not a Pastoral poet. Other peasant poets lose their innocence by being seduced by the demands of formal diction, but Clare resists this; his voice is the oaten reed itself, not the sound of the Pastoral Pipe overheard by a poet who has learned to orchestrate it and play antiphonal effects with it. If this cultural and social antithesis is absent, so too is the chronological antithesis, for Clare's best work is dedicated to the here and now, 'Clare at his least dilute preferred the *is* of everything he celebrated.'[8] It is equally inaccurate to call Clare a georgic poet. Although his work shares the abundant natural detail of georgic verse from Hesiod and Virgil to Christopher Smart, it cannot be said to consist, as Addison puts it, 'in giving plain and direct instructions to the reader ...'.[9] Yet one can approach Clare through georgic poems in one important respect: his people, like theirs, are at work in the landscape, not merely enjoying it or idealizing it.

The elimination of the realities of work is a characteristic of Renaissance and neo-classical Pastoral and can be compared with the harsher antitheses of Theocritus and Virgil. Clare's work has the vivid realism of the early pastoral poets, made more urgent and personal by a Romantic temperament *in extremis*. Inasmuch as Clare is a poet, one of the 'abstracts and brief chronicles of the time', he is distanced from his own background. His landscape is very different from most other literary landscapes, but it is necessarily the result of the observation of a unique sensibility, and the isolation which that entailed. Clare has, on the one hand, the suspicious glances of his fellow-villagers, and on the other, the officious supervision of his editors. His position is very rare: close enough to the soil to experience its cruel intractability and joyous tang, he is sufficiently distanced to see his environment as a whole.

As he becomes aware of his isolation, Clare makes a unique contribution to Pastoral by virtue of his extreme sadness. For the Eden Clare lost was not only a childhood as vivid and sensuous as Wordsworth's, but also one which was rooted in a landscape which itself disappeared. It is the sadness and anger of a whole people that Clare articulates. Much Pastoral depends upon the image of a Golden Age at some unspecified time in the past, but Clare lost an inner and outer Eden together, so that he can employ the convention of nostalgia with unusually exacting poignancy. The leaden present and the golden past are not just historical or mythological, they are personal experiences.

If Clare is not a 'pastoral', 'nature', 'landscape', or 'topographical' poet in any of the accepted senses of these terms, he can more accurately be called a 'picturesque' poet, and I have found it fruitful to approach his work in relation to the Picturesque. The best summary of the Picturesque is given in a long note to the Introduction of *The Romantic Agony* by Mario Praz (London, 1970). He begins:

The 'picturesque' (a word, as is well known, of Italian origin, meaning the point of view essentially of a painter) was elaborated as a theory in

England between 1730 and 1830 ('Le Pittoresque nous vient d'Angleterre'--Stendhal) and was, in a sense, a prelude to Romanticism. If we regard the 'picturesque' period merely as the period during which this particular point of view became conscious and assumed the importance of a fashion, then undoubtedly the English poets Thomson and Dyer, with their descriptions which translate into terms of literature the pictorial manner of Claude Lorrain and Salvator Rosa, are the godfathers of the Picturesque. (p. 19)

If Thomson is one of the godfathers of the Picturesque, then Clare is one of his godsons (as are Constable and Turner). Clare was steeped in Thomson's work, and he absorbed the picturesque technique of vision before adapting it to his own needs. The characteristics of the Picturesque listed by Praz are basic ingredients of Clare's work: 'dazzle and flicker of effect, ... rapid succession of colours, lights, and shades, ... roughness, sudden variation, irregularity.' (Ibid., p. 21.)

To these kinetic qualities Clare adds a keen eye for minute detail. The mobile and the static are fused together, as if his eye had a special lens, registering effects of movement not fully captured until the invention of the cine-camera. By 1841, when Clare entered the Asylum (for the second and final time), Victorian art had begun to move away from the joyous interchange of natural and human powers celebrated in the best Romantic art. Much Victorian art seems frozen in a kind of hyper-sensitive stasis. In *The Finer Optic: The Aesthetic of Particularity in Victorian Poetry* (Yale, 1975), Carol T. Christ comments tellingly on this:

Millais's picture of Ophelia drowning amid flowers that dazzle the spectator stamen by stamen [has] a microscopic exactitude artificial to normal vision, almost as if a whole generation of artists were born near-sighted. This peculiar consciousness of the minute particulars of nature furthermore plays a paradoxical role in the work of these artists. Although they pursued the accurate observation of nature more zealously than any Romantic, they decreasingly felt any sympathetic power in nature. The uses of acute sensitivity to natural detail thus often become curiously detached from any interest in nature itself. (p. 18)

It is as if Clare, who in the period dealt with in this book kept inner and outer fused in a precarious balance, loses his reason before the encroachment of too much detail. As Carol Christ puts it, when we come across minute detail in Victorian art or poetry, it is often a sign of obsession, madness, or of 'a character fixated in some morbid emotion' (*The Finer Optic*, p. 12). And to a modern taste, the atmosphere of a Victorian drawing-room seems to stifle the onlooker with objects.

This feeling of helplessness in a world of increasing particularity and atomism is analysed as early as 1882 by Frederic Harrison in an article called 'A Few Words about the Nineteenth Century':

A hundred years ago, a naturalist was a man who, having mastered, say, some millions of observations, had, if he possessed a mind of vigour, some idea of what Nature is. Now, there are millions of billions of possible observations, all in many different sciences, and as no human brain can deal with them, men mark off a small plot, stick up a notice to warn off intruders, and grub for observations there.[10]

Clare, born into the world of Gilbert White, survived into the age of Darwin. His best work between 1821 and 1841 has the innocent absorption of the former and the awareness of change and adaptation of the latter. He had mastered 'some millions of observations' and was able to incorporate this knowledge into work which, more than that of any other Romantic, challenges the neo-classical demand for the general and the ideal. This work has a wholeness which appeals directly to the modern reader, isolated as he is in his 'small plot'.

In his book *Picturesque Landscape and English Romantic Poetry* (London, 1970), J. R. Watson shows how Wordsworth, Coleridge, Shelley, Scott, Byron, and Keats assimilate the picturesque tradition, making it an ingredient in their own more comprehensive visions. My book sets out to show what Clare does with this tradition. The chief difficulty that presents itself in this approach is that Clare does not document his views as fully as the others, especially Wordsworth. I do not know if Clare had read Gilpin, or Uvedale Price, or Payne Knight, as we know that Wordsworth certainly had. Although these three authors define the Picturesque in different ways, it is the common elements of sensibility implicit in the term, as listed by Praz above, which concern us.[11]

The study of Clare has its peculiar pleasures, and its peculiar difficulties. One of the difficulties, paradoxically, arises from Clare's surface simplicity. Philip Larkin explains a similar difficulty in his discussion of the criticism of Thomas Hardy's poetry. In the present context, read 'Clare' for 'Hardy':

We can say that modern criticism thrives on the difficult ... and that Hardy is simple; his work contains little in thought or reference that needs elucidation, his language is unambiguous, his themes readily comprehensible. A typical role of the modern critic is to demonstrate that the author has said something other than he intended ... but when this is tried on Hardy ... the reader feels uncomfortable rather than illuminated ...[12]

In the belief that Larkin's words can apply equally to Clare, I have avoided an overemphasis on purely textual analysis. That said, however, I believe that a full understanding of Clare's achievement is by no means easy, and that the situation is not made any easier by the fact that few critics have given him their full attention, and fewer still have seen him against the wider context which can cast light upon the defeats and triumphs of his work. In the final analysis, there is no poet with whom Clare can be compared, so irrevocable is his expulsion from Eden, so complete is his isolation, so unbroken is his spirit. He stands in our tradition as he stood in life, 'the Crusoe of his lonely fields'.[13]

Clare's spelling and grammar are often unorthodox, but with a little patience, the reader will soon become accustomed to his language. When an expression or dialect word is likely to be obscure, it has been glossed, but annotation of this kind has been kept to a minimum. Clare's absence of punctuation should also cause little difficulty. The quotations from Clare's poetry have been taken, as far as possible, from *Selected Poems and Prose of John Clare* (Oxford, 1967; Oxford Paperback, 1978) and *The Shepherd's Calendar* (Oxford, 1964; Oxford Paperback, 1973), both edited by Eric Robinson and Geoffrey Summerfield. But these contain only a fraction of Clare's pre-Asylum work. When a poem is not in these editions, I have quoted from *The Midsummer Cushion* (Mid Northumberland Arts Group, 1979)[14]. On just one occasion, I have quoted from the J. W. Tibble edition (Dent,

1935), in order to make a critical point. Valuable as this edition
is, it is an unsatisfactory source, as Tibble 'improved' many of
the manuscripts, altering spelling, punctuation, and grammar.
Most criticism since 1935 has had to rely on this edition; it is my
conviction that Clare's work loses a great deal by being 'improved'.[15] In the case of poems where there is no reliable printed text,
I have quoted either from the forthcoming Oxford Poets Clare,
edited by Eric Robinson and David Powell, or from the unedited
transcripts which will form the basis of future Oxford English
Texts editions, also being prepared by Eric Robinson and David
Powell.

1

'Picturesque Green Molehills'

I

For many readers of poetry, John Clare (1793-1864) is the epitome of the Romantic poet, and there undoubtedly is a romantic appeal in the story of his triumph over almost insuperable odds. The story of his birth in a labourer's cottage, his brief rise to fame after the publication of his first two volumes, his struggles with poverty, neglect, and condescension, his long walk of escape from the first Asylum and his final incarceration at Northampton, is familiar enough to most readers. But we must not let these biographical facts obscure the most important story of all, that of Clare's development as an artist. In this chapter, and throughout this book, I examine his struggles to find his own voice within the topographical tradition which he inherited from the eighteenth century. Out of such material, Clare creates a unique vision, in which detail is involved in an animated environment as whole and mysterious as that of any other Romantic poet or painter.

Clare's dislike of formality—whether it took the shape of landscape gardening, punctuation, or traditional verse-form—should not blind us to the fact that his work as a whole shows considerable technical virtuosity and variety. Like every original artist, he was steeped in the tradition against which he rebelled. He had a delicate ear which absorbed every nuance of rhythm from his wide reading; hundreds of his lines have a memorable felicity. If it is fair to say that he often fails to achieve a completely harmonious poem (felicity has its drawbacks, as line crowds upon line), it must also be acknowledged that his successes are numerous. Readers of the recently-published *The Midsummer Cushion* will see at once that Clare's technical ability was of a high order and that he never ceased to experiment; a 'self creating joy'[1] was for him pre-eminently the gift of poetry. If this volume had not lain in manuscript for nearly 150 years, our view of Clare today might be very different.

Before proceeding, let us look briefly at a little-known poem, 'Evening Primrose', which seems to me to achieve a minor perfection. This poem has been set to music by Benjamin Britten:

> When once the sun sinks in the west
> & dewdrops pearl the evenings breast
> All most as pale as moonbeams are
> Or its companionable star
> The evening primrose opes anew
> Its delicate blossoms to the dew
> & shunning-hermit of the light
> Wastes its fair bloom upon the night
> Who blindfold to its fond caresses
> Knows not the beauty it posseses
> Thus it blooms on till night is bye
> & day looks out with open eye
> Bashed at the gaze it cannot shun
> It faints & withers & is done
>
> *(MC 432)*

Clare places this poem among the sonnets in *The Midsummer Cushion*; it has fourteen lines, but it is highly unusual, even for Clare, in that it is written in octosyllabic couplets (Clare sometimes writes sonnets in decasyllabic couplets). This form is suitable for the medium-length discursive or descriptive poem, and Clare uses it skilfully on many occasions. There is a shock of delight in seeing it used here so sparingly—as in an Oriental water-colour, one brush-stroke too many would spoil the delicacy of this piece.

One can notice here Clare's use of personification, his use of the literary word 'opes' and the strong echo of Gray's Elegy in line 8. Yet who else but Clare could have written line 7, '& shunning-hermit of the light', with its distinctive and poignant compound noun? One is tempted to echo Coleridge's praise of Wordsworth and say that if one heard this line in the desert, one would shout out 'Clare'. And how audaciously exact is Clare's use of 'companionable', which reminds one of Yeats's equally effective use of this word. In the Elegy, Gray strikingly anticipates the Romantics in his creation of a brooding ghost-like figure (later used so hauntingly by Hardy), whose presence colours our reaction to the whole landscape. The experience of reading Clare is somewhat different; Clare has a Chaucerian ability to present the bright details of nature and be moved by them while remaining, as author, very much in the background. In a poem such as 'Evening Primrose', instead of a Wordsworthian 'egotistical sublime', we get a humble ability to let nature speak for itself. Clare knows, of course, that nature cannot speak entirely for itself, but he is determined not to give it lessons in elocution. In the following pages, I attempt to elucidate the process whereby Clare assimilates his literary heritage and, by virtue of his unusual perspective, creates a body of poetry of considerable originality.

II

The Romantics prided themselves on looking at nature face to face. This way of seeing was made possible, however, by the increasing fascination with the mechanics of vision during the eighteenth century. The development of optical instruments is an intriguing foretaste of the camera, and ultimately the cinema. By 1800 there was a profusion of such instruments, including the magic lantern, the Claude glass, the *camera obscura*, the *camera lucida*[2] and 'a whole family of panoramas, dioramas and other offspring, among them Thomas Girtin's Eidometropolis, a panorama of London ...'.[3] Perhaps the most interesting of all was de Loutherbourg's Eidophusikon 'in which moving pictures were shown within a proscenium, by means of a combination of Argand lamps, coloured gauzes, lacquered glass and receding planes, which reproduced scenes with realistic atmospheric effects.' (Ibid., p. 141.) The Eidophusikon was first shown in 1781 and was admired by Gainsborough and Reynolds. With its stress on the kinetic and the atmospheric, the Eidophusikon was hardly a model for the neo-classicist, but Reynolds, as in certain celebrated passages in his *Discourses on Art*, is fascinated beyond the confines of his own theories. In one such passage, discussing Gainsborough's 'minute observation of particular nature', Reynolds writes, 'If Gainsborough did not look at nature with a poet's eye, it must be acknowledged that he saw her with the eye of a painter; and gave a faithful, if not a poetical, representation of what he had before him.'[4] This neatly summarizes the transition which occurred in this period from a view of the painter as one who purveys generalized knowledge and morality to one whose business it is to pursue the painterly, tactile nuances of his medium.

Gilpin, similarly, is drawn beyond his more rational self by a fascination with optical processes. The usual view of Gilpin is of a rather old-maidish clergyman with a taste for scenery, and sometimes he does nothing to dispel this view; he can dismiss certain landscapes as 'disgusting' for not living up to his rules of composition. But scattered about his work, there are passages in which he becomes so interested by the unique visual qualities of a scene that he delineates an autonomous, dynamic landscape. In his *Observations on the River Wye* (London, 1782), there are several passages in which he uses his eye like a cine-camera, as his boat moves down the river, or as the 'shifting' scenes 'float' past his carriage window. But he is unable to invent a verbal or pictorial technique for these visual sensations:

Many of the objects, which had floated so rapidly past us, if we had had time to examine them, would have given us sublime, and beautiful hints in landscape: some of them seemed even well combined, and ready prepared for the pencil: but, in so quick a succession, one blotted out another. (p. 72)

Gilpin also used the Claude glass, a tinted convex mirror often carried by tourists and artists, which gave the landscape the ideal qualities of a Claude painting. He describes the effects of using 'the mirror' while travelling in his coach, showing a persistent attempt to transcend the limitations of a primitive instrument:

They are like the visions of the imagination; or the brilliant landscapes of a dream. Forms, and colours in brightest array, fleet before us; and if the transient glance of a good composition happen to unite with them, we should give any price to fix and appropriate the scene.[5]

Thomas Gray also used a Claude glass. On 2 October 1769, while touring the Lakes, he wrote, '... fell down on my back across a dirty lane with my glass open in one hand, but broke only my knuckles: stay'd nevertheless, & saw the sun set in all its glory.'[6]

It is unlikely that Clare had read Gilpin, but Gray's letters were favourites of his. As we shall see later in this chapter, Clare had personal reasons for disliking optical instruments. As an artist, he casts aside, as it were, the Claude glass (whose user had to turn his back to the landscape), and involves himself directly with sensuous elements, eventually creating a verbal equivalent for the 'shifting' and 'floating' scene. In the following passage from 'The Autobiography', he testily dismisses the artificiality of 'an instrument from a shilling art of painting' (obviously a primitive *camera obscura*), in favour of the 'instantaneous sketches' of living nature. It is a dismissal as conscious, radical, and intelligent as Constable's rejection of the academic 'brown tree':

Sometimes he woud be after drawing by perspective & he made an instrument from a shilling art of painting which he had fashiond that was to take landscapes almost by itself it was of a long square shape with a hole at one end to look through & a number of different colourd threads crossd into little squares at the other from each of these squares different portions of the landscape was to be taken one after the other & put down in a facsimile of the square done with a pencil on the paper but his attempts made but poor reflections of the objects & when they were finishd in his best colours they were but poor shadows of the original & the sun with its instantaneous sketches made better figures of the objects in their shadows[7]

One can gather from these lines that Clare disagrees in principle with this machine with its fussy little squares and its 'drawing by perspective'. He reacts by invoking the aid of a much greater artist, 'the sun with its instantaneous sketches'. Man, by comparison, is reduced to sketching a 'facsimile'. Clare's ambition as an artist is to be alive to what Constable called the 'CHIAR'-OSCURO OF NATURE', and to vie with the sun in producing 'instantaneous sketches' of nature's scenes.

The forces of nature are one thing, however, and the mind of man another, a distinction which Clare's personification helps to obscure. Some sort of filtering and ordering process must go on to produce consciousness and especially that heightened aspect of consciousness which produces art. Even the simplest response to landscape, as Gilpin is aware, involves more than the flickering, evanescent peep-show granted to him in his moving carriage. For the mind is not just a screen over which fleeting images pass, the eye is not just a mirror-lens, and when one takes a walk, one is more than, in Stendhal's phrase, 'a mirror, carried along a roadway'. When Hamlet advised the players to 'hold as 'twere the mirror up to nature', he meant that the actor should be able to imitate the generalized pattern of human behaviour; something very different happened when Thomas Gray took a walk in the Lake District with his Claude glass. When Pope wrote in *An Essay on Criticism*, 'First follow NATURE, and your Judgment frame/ By her just Standard, which is still the same',[8] he was invoking the essential harmony of the cosmos as a model for human nature. But Pope, in his capacity as landscape gardener, was also one of the chief pioneers of that change of taste in the eighteenth century whereby Nature in its neo-classical sense became landscape or scenery—what Wordsworth was to worship, 'Knowing that Nature never did betray/ The heart that loved her'.[9]

It is important to remember how recently the concept of nature as scenery evolved. Nature in the medieval world existed as a decorative backdrop or as a narrative or moral device. The word 'landscape' did not emerge until the late sixteenth century. Milton's 'L'Allegro', with its 'Lantskip', is dated 1631 or 1632. The growing rationalistic climate of the seventeenth century exploited the Renaissance systematization of perspective as a 'natural' way of seeing things, and neo-classical landscape reached a high degree of sophistication in Claude and Poussin. John White writes:

During the thirteenth and fourteenth centuries it was possible to see space gradually extending outwards from the nucleus of the individual solid object. ...Now the pictorial process is complete. Space is created first, and then the solid objects of the pictured world are arranged within it in accordance with the rules which it dictates. Space now contains the objects by which formerly it was created.[10]

New attitudes to landscape, from the mid-seventeenth century onwards, began to demand more than a spatial box in which objects could be placed and framed. In garden design, for example, the formal parterres and geometrical lay-out borrowed from the French gave way in the late seventeenth century to a freer, more informal style, epitomized by such figures as Sir William Temple, Addison, Pope, and William Kent. The theory and practice of such men led to what was known all over Europe as 'the English garden'. To walk around the gardens of Stowe, Stourhead, or Rousham in their prime was to be involved in landscape and in time in new ways, and to perceive new relationships between space and time. The picturesque vistas were modified as one walked, or one could pause to compose a formal picture, or abstract a moral, or meditate upon a *memento mori* or *sic transit gloria mundi* theme. For men such as Pope and Kent and Shenstone, this was a process of self-discovery as much as it was a merely associative game. They took it as axiomatic that the training of the eye was a moral activity, in that a properly conceived, and perceived, landscape or garden was an emblem of order (but not regimentation), in the state, the mind, the soul, and the emotions. This was the assumption behind all topographical poetry from Denham's 'Cooper's Hill' onwards. The balance of the couplet, with its suitability for antithesis and irony and its careful punctuation, which must be respected like dynamics in music for its full effect, is the ideal vehicle of this moral fear of extremes. Denham apostrophizes the Thames in a passage which illustrates this:

> O could I flow like thee, and make thy stream
> My great example, as it is my theme!
> Though deep, yet clear, though gentle, yet not dull,
> Strong without rage, without ore-flowing full.[11]

The French- or Dutch-style gardens in England can be seen in the bird's-eye views of country seats in Kip and Knyff's *Britannia Illustrata* (1707 *et seq.*), in which the landscape is severely regimented almost to the horizon. This style carries out the process in which 'Space is created first, and then the solid objects of the pictured world are arranged within it in accordance with the rules which it dictates.' The well-ordered walks with segments of lawn, topiary hedges, and strategically placed statues or urns are like so many corridors, rooms, and pieces of furniture. Marvell's Mower had described such a garden, 'He first enclos'd within the Garden's square/ A dead and standing pool of Air.'[12] Such a passage as the following, from Shaftesbury, had a profound influence on the new attitudes:

I shall no longer resist the passion in me for things of a natural kind; where neither Art, nor the Conceit or Caprice of Man has spoil'd their genuine order, by breaking in upon that primitive State. Even the rude Rocks, the mossy Caverns, the irregular unwrought Grotto's, and broken Falls of Waters, with all the horrid Graces of the Wilderness itself, as representing NATURE more, will be the more engaging, and appear with the Magnificence beyond the formal Mockery of princely Gardens.[13]

Picturesque gardeners were agents in a process whereby space began to be conceived in new ways. The Romantics continued this process; they unframed nature just as they broke down or discarded traditional forms. They inherited the picturesque way of looking at nature, but realized that it, in its turn, had become a tyranny, so they invented new ways of seeing which were new ways of feeling. Landscape became Moodscape, dictated its own terms and created its own space. The Romantics experienced an intimacy with landscape, with which they were involved as with a lover or friend. Wordsworth, who was a devoted and experienced gardener,[14] saw the intellect as 'wedded to this godly universe/ In love and holy passion' (*Wordsworth: PW* v. 4); and Coleridge wrote, 'O Lady! we receive but what we give,/ And in our life alone does Nature live:/ Ours is her wedding garment, ours her shroud!' (*Poetical Works* (London, 1969), p. 365.)

This intimacy precluded seeing nature from the outside, or *de haut en bas*, as in prospect. It required involvement, commitment, the ability to see through the eye, not with it. To call Clare a Romantic is usually misleading, but inasmuch as he fundamentally mistrusts any preconceived design or framework, including traditional verse-forms and punctuation, Clare is Romantic. In Clare's world, space gradually extends 'outwards from the nucleus of the individual solid object'. This makes it very different, not only from neo-classical theory, with its love of everything in its 'right' place, but also from the theories of such men as Gilpin, Price, and Repton. Picturesque gardening, painting, and architecture, for all their 'Romantic' feelings for the landscape, were still fundamentally guided by what Wordsworth called the 'meddling intellect'. Gilpin writes that the task of the picturesque eye is to *survey* nature; not to *anatomise* matter. It throws its glance around in the broadcast stile. It comprehends an extensive tract at each sweep. It examines parts, but never descends to particles.'[15] In his respect for the general, Gilpin is nearer to Johnson than to Coleridge, let alone to Clare, whose universe is entirely made up of particles. If Gilpin anticipates the Romantics in his feeling for the kinetic, when it comes to the microscopic he remains in the neo-classical camp.

It is not difficult to observe qualities of 'meddling intellect' in the following passage by Humphry Repton, which accompanies a 'before' and 'after' plate. Repton is explaining how he has 'improved' the view from his cottage window:

... the humble Cottage ... stood originally within five yards of a broad part of the high road: this area was often covered with droves of cattle, of pigs, or geese. I obtained leave to remove the paling twenty yards farther from the windows; and by this *Appropriation* of twenty-five yards of Garden, I have obtained a frame to my Landscape; the frame is composed of flowering shrubs and evergreens; beyond which are seen the cheerful village, the high road, and that constant moving scene, which I would not exchange for any of the lonely parks, that I have improved for others.[16]

The consequence of this ingenious form of enclosure is that the cattle, pigs, geese, butcher's shop, stage-coach, and one-armed beggar leaning over the fence all disappear when Repton pulls down the flap of his 'after' plate. The rustic scene at once becomes suburbia.

Clare, in writing about enclosure, describes the fencing in of 'little parcels little minds to please'.[17] Another view from the wrong side of Repton's fence was obtained by the Eskimos brought to England by the Labrador explorer, Captain Cart-wright. 'The land is all made' was their comment on the English landscape.[18] Horace Walpole, looking out from his carriage window, was of a similar opinion: 'The journey is made through a succession of pictures.'[19] This remark implies that the educated person, like the young ladies in Jane Austen, was expected to be visually accomplished. These three quotations give some idea of the extent to which the face of the landscape was being changed for agricultural, economic, aesthetic, or territorial purposes. W. G. Hoskins, in *The Making of the English Landscape* (1955) dispelled the notion that the landscape was primarily the creation of aristocratic landscapists and parliamentary enclosure, but to this day, a large amount of English landscape initially gives that impression. Regional variety must be allowed for, of course. The Eskimos and Walpole were seeing the more 'cultivated' counties, in both senses of the word. But even remote Helpston was transformed by parliamentary enclosure within Clare's lifetime, and great landscaped parks existed within walking distance. As we will see in Chapter Four, Clare knew the servants at Milton, a park improved by Repton, and was a gardener's assistant for a short while at Burghley, which Capability Brown had improved.

In a crucial passage of *The Prelude*, Wordsworth condemns this process of selecting and comparing picturesque images. He writes that this

> ... strong infection of the age,
> Was never much my habit—giving way
> To a comparison of scene with scene,
> Bent overmuch on superficial things,
> Pampering myself with meagre novelties
> Of colour and proportion ...[20]

This was never much Clare's habit, either, and he had more immediate reasons to be suspicious of it than Wordsworth. They are both attentive 'to the moods/ Of time and season ... and the spirit of the place' (ibid.). Yet Clare, like Wordsworth, was highly sensitive to colour and proportion. They both loved the art of painting, gave much thought to it, and found great pleasure in the friendship of painters. Wordsworth, as J. R. Watson has shown, constructs the opening of 'Tintern Abbey' in a painterly fashion, with three planes of distance.[21] What makes the poem great is the process whereby, in Coleridge's words, Wordsworth 'dissolves, dissipates and diffuses' that picturesque scene, 'in order to recreate' it.[22] This is the Imagination at work. Similarly, Clare's poetry is full of painterly impressions and sketches of attractive spots, and he often uses the adjective 'picturesque'. Belonging to the second generation of Romantics, Clare would have noticed how Wordsworth and Coleridge transform the framework of most eighteenth-century poetry and allow objects and characters to create their own space or their own motion. Listen to the spacing of Wordsworth's 'Influence of Natural Objects':

> And not a voice was idle: with the din
> Smitten, the precipices rang aloud;
> The leafless trees and every icy crag
> Tinkled like iron; while far-distant hills
> Into the tumult sent an alien sound
> Of melancholy, not unnoticed while the stars,
> Eastward, were sparkling clear, and in the west
> The orange sky of evening died away.
>
> (*Wordsworth: PW* I. 249)

In 'This Lime-Tree Bower my Prison' (1797), Coleridge imagines his friends, on their walk, pausing to look over a picturesque scene:

> Now, my friends emerge
> Beneath the wide wide Heaven—and view again
> The many-steepled tract magnificent
> Of hilly fields and meadows, and the sea,
> With some fair bark, perhaps, whose sails light up
> The slip of smooth clear blue betwixt two Isles
> Of purple shadow!
>
> (*Poetical Works* 179)

At the end of the poem, a single rook can eliminate that well-spaced prospect with the stereophonic effect of its 'creeking' flight:

My gentle-hearted Charles! when the last rook
Beat its straight path along the dusky air
Homewards, I blest it! deeming its black wing
(Now a dim speck, now vanishing in light)
Had cross'd the mighty Orb's dilated glory,
While thou stood'st gazing; or, when all was still,
Flew creeking o'er thy head, and had a charm
For thee, my gentle-hearted Charles, to whom
No sound is dissonant which tells of Life.

(Ibid. 181)

The next section of this chapter shows how Clare transforms his own picturesque vision by examining one prolific branch of topographical poetry, the prospect or hill poem.

III

Topographical poetry in the wide sense is as old as poetry itself, for poets have always felt the need to celebrate their environment, thereby giving their emotions 'a local habitation and a name'. Drayton's *Poly-Olbion* and Jonson's 'To Penshurst' are important poems in this tradition. But the publication in 1642 of Sir John Denham's 'Cooper's Hill' inaugurated a special type of topographical poem which had its full flowering in the eighteenth century and which was, in its decline, part of the Romantic poets' inheritance. In this tradition, analogous to the pictorial tradition of the birds's-eye view, the poet climbs a hill, sweeps his eye round the panorama, focuses his attention on the most interesting prospect, and paints a careful word-picture.[23] It did not need much of an effort for eighteenth-century minds to associate the hill in question with Parnassus, but it was also Denham who had domesticated Parnassus for them, as he makes clear in the first eight lines of 'Cooper's Hill':

Sure there are Poets which did never dream
Upon *Parnassus*, nor did tast the stream
Of *Helicon*, we therefore may suppose
Those made not Poets, but the Poets those.
And as Courts make not Kings, but Kings the Court,
So where the Muses & their train resort,
Parnassus stands; if I can be to thee
A Poet, thou *Parnassus* art to me.[24]

Topographical poetry is a branch of landscape poetry and sets out, in R. A. Aubin's words, to describe *'specifically named actual localities'*.[25] Prospect poetry is a specialized type of topographical poetry in which an extensive view is described from one or more points (or 'stations') and in which the poet draws moral or patriotic lessons from the scenery. The ambiguity in the word 'prospect' is always present—a 'prospect' is a long view both in space and in time. While Aubin sees the essential characteristics of the topographical genre as those which name and describe specific places, John Wilson Foster has attempted to define the genre more precisely.[26] According to Foster, in topographical poetry certain structural devices are employed to describe the landscape; the genre can be recognized not only by its naming and description of specific places, but by the way in which description is controlled by spatial, temporal, and moralistic designs, 'We can say broadly that topographical poetry ceases to warrant the title when the poet no longer sets up certain kinds of descriptive patterns to convey corresponding patterns of moralistic meditation.' (Ibid., p. 404.)[27]

Foster elsewhere stresses the connection between topographical poetry and the sciences of surveying and topography.[28] In discussing 'Cooper's Hill', he writes. 'The eye-shifts in "Cooper's Hill" bear a close resemblance to the way sightings were taken with these [surveyors'] instruments ...' (ibid., p. 242). He explains in another article that enclosure became a socio-economic spur to the development of surveying.[29] It follows that for Clare the word 'survey' carries a double threat. On the one hand, scientific surveyors were enclosing the landscape of his boyhood; on the other, topographical artists and poets 'overlooked' what made his life and landscape meaningful. The enclosure of the landscape around his Northamptonshire village, Helpston, proceeded from 1809; later he recorded his reactions to surveyors at work:

Saw 3 fellows at the end of Royce Wood who I found were laying out the plan for an 'Iron railway' from Manchester to London it is to cross over Round Oak Spring by Royce Wood corner for Woodcroft Castle I little thought that fresh intrusions would interrupt & spoil my solitudes after the Enclosure they will despoil a boggy place that is famous for Orchises at Royce Wood end (*Prose* 151)

Enclosure, landscape gardening on a huge scale, and the creation of the railways swept away between them much of that open nature whose loss Clare continually laments. The changes were carried out by a horde of surveyors, representatives of what John Barrell calls the 'rural professional class',[30] with little sympathy for or knowledge of Clare's way of life.

Foster's analysis of structural devices is especially applicable to the sub-genre of prospect poetry. If one rests content with Aubin's definition, the nature of Clare's resistance to the tradition is unclear; if one sees topographical poetry as merely naming and describing specific places, then Clare could be described as a topographical poet, for many of his poems do just that. But once the connection with scientific surveying is made explicit, Clare's predicament becomes clearer. It is perhaps a sense of threat, as much as genuine literary influence, which attracted Clare to landscape poetry before Denham (or outside the Denham tradition). Clare's affinities with the seventeenth century are evident from his skilful exercises, 'fathered' upon Sir Henry Wotton, Sir John Harington, Andrew Marvell, and others, grouped together by J. W. Tibble under the heading 'Poems Written in the Manner of the Older Poets', and dated between 1824 and 1832.[31]

Clare's approach to eighteenth-century poetry, however, was necessarily more complex. He would have noticed how optical and landscape terms had become assimilated into political and social metaphors—'viewpoint' or 'point of view' is an intellectual term, one has 'elevated' thoughts by being in an 'elevated' position, one's life gains 'perspective' as well as the landscape (painted or real), one should accept one's 'walk' of life, one ought not to have ideas above one's 'station', one's life has 'landmarks' if one 'surveys' it properly, and to 'command' bright 'prospects' is more than just having mental snapshots of the view, or 'panorama', from an 'eminence', preferably from a 'seat' (garden seat or family seat). Clare's problem as an artist is how to write descriptive poetry about his own landscape without recourse to this alien vocabulary.

If Denham's view from Cooper's Hill could be called telescopic (it is far-ranging but limited to one direction at a time), Clare's vision could be called kaleidoscopic (it is not concerned with distancing but with comprehensiveness, a circular all-at-onceness).[32] Apart from juvenilia and occasional pieces for the Annuals, his vision is never framed and rarely static, he has no conventional point of view in either sense, his thoughts are hardly ever elevated, he eschews perspective, the landmarks of his life are nearly all records of loss, and he had, in both senses of the word, few prospects. He comes to maturity by discovering that all things, even the snails which he had timed with his watch, are in motion, and that objects are often best seen in close-up, thereby achieving a kind of fluid crystallization of images.

No wonder Clare was nervous when his critics used topographical metaphors. For him they were the language of *dis*integration, *dis*location, *dis*orientation. His publishers, Taylor and Hessey, who vigorously corrected his manuscripts against his will, wrote to him with advice such as, '... if you would raise your views generally & Speak of the Appearances of Nature each Month more philosophically ...' and 'What you ought to do is to elevate your Views, and write with the Power that belongs to you under the Influence of true Poetic Excitement—never in a low or familiar Manner ...' and 'let [your descriptions] come in incidentally—let them occupy their places in the picture, but they must be subordinate to higher objects.'[33] It is a great tribute to Clare as an artist that he ultimately refused to listen to this sort of advice; his refusal made him a poet, but it destroyed his material prospects. His finest book, *The Shepherd's Calendar*, which appeared in 1827 after years of tampering by John Taylor, fell on deaf ears, and an accurate edition did not appear until 1964.[34]

The prospect formula is part of the neo-classical belief in the truth of the generalized, the idealized, the elevated. Dr Johnson, discussing the pastoral genre, writes: '... though nature itself, philosophically considered, be inexhaustible, yet its general effects on the eye and on the ear are uniform, and incapable of much variety of description'.[35] Clare's poetry is a direct challenge to that statement, as it is to Sir Joshua Reynolds's dictum, 'All smaller things, however perfect in their way, are to be sacrificed without mercy to the greater.' (*Discourses on Art*, p. 56.) Clare's response is in the spirit of his age—botanists were making a vast number of discoveries at this time, which were then codified and

illustrated. Following the pioneering work of Linnaeus, important English botanical works included Thornton's *Temple of
Flora* (1799-1807), Curtis's *Flora Londinensis* (1777-87), and
James Sowerby's *English Botany*, which ran to 36 volumes
between 1790 and 1814. Clare himself was a skilful and knowl-
edgeable botanist as well as ornithologist, and his library of about
440 volumes, still in existence at Northampton, contains many
books on natural history. Inside the front cover of his copy of
Isaac Emmerton's *The Culture & Management of the Auricula,
Polyanthus, Carnation, Pink, and the Ranunculus* (1819), Clare
has written a list of 22 'Orchis's counted from Privet hedge'.
Through his friend Henderson, the butler of his patron Earl
Fitzwilliam at Milton House, Clare occasionally got glimpses of
finer things. In the following extract from the Journal of 15 Dec-
ember 1824, Clare's instincts run directly counter to Reynolds's
generalization, as he carefully checks a personal discovery against
the latest scientific codification:

Went to Milton saw a fine edition of Linnaeus's Botany with beautiful
plates & find that my fern which I found in Harrisons close dyke by the
wood lane is the thorn-pointed fern saw also a beautiful book on insects
with the plants they feed on by Curtis (*Prose* 127)

The higher the viewpoint, the more generalized and idealized
the view; the lower the viewpoint, the more particular details
will crowd out or even obscure the general and ideal, and assert
their own disturbing independence. It is no accident that Dutch
landscape painting, given low priority by Reynolds in the
Discourses on Art, was admired by English Romantic painters, including Clare's favourite painter, De Wint, whereas the favourite
landscape painter of the eighteenth century was Claude Lorrain,
whose tradition Clare detests. Nor is it accidental that the first
indigenous school of English water-colours emerged in Norwich,
in the midst of the great East Anglian plain, where the painter,
having few conventional prospects, is forced into a direct contact
with the atmospherics of the scene under a huge sky. Gainsborough in his early work such as 'Cornard Wood', and Constable in his most typical work, were both inspired by the Suffolk
landscape. The bridge between the Claudean bird's-eye view

(used by Dyer and Thomson) and an increasingly lowered view-point which demands attention to detail is to be found in the development of the Picturesque. At the beginning of the century, poets learned to orientate themselves in landscapes by adapting the Claudean view to verse; by 1800, the cult of the visual had led to a complex awareness of stimuli, so that lichens or moss on old stonework could provide not only synaesthetic emotion ('The sight takes so many lessons from the touch' wrote Uvedale Price to Sir George Beaumont)[36] but also could lead the observer into a magically microscopic world, where pictorial rules became irrelevant. Wordsworth writes of the Alps that they abound in 'images which disdain the pencil' and he transcends the Picturesque by experiencing the Sublime;[37] Clare, as visually acute as Wordsworth, transcends the Picturesque by discovering an equally awe-inspiring microcosmos.

The neo-classical theorist maintains that the poet does not number the streaks of the tulip, but Clare's eye level is often no higher than a tulip, and he numbers meticulously:

> With the odd number five strange natures laws
> Plays many freaks nor once mistakes the cause
> And in the cowslap peeps* this very day
> Five spots appear which time neer wears away
> Nor once mistakes the counting—look within
> Each peep and five nor more nor less is seen
> And trailing bindweed with its pinky cup
> Five lines of paler hue goes streaking up
> And birds a many keep the rule alive
> And lay five eggs nor more nor less then five
> (*SPP* 111)

Pope's well-known passage from *An Essay on Man* reads like an accusation of Clare's passionate botanizing:

> Why has not Man a microscopic eye?
> For this plain reason, Man is not a Fly.
> Say what the use, were finer optics giv'n,
> T'inspect a mite, not comprehend the heav'n?
> (*Pope: Poems* III. i. 38-9)

But Clare saw no reason why man should not have such a vision. His curiosity is scientific as much as poetic; indeed the modern separation of these two faculties would have been meaningless to him. He carefully records what he calls 'snatches of sunshine and scraps of spring that I have gathered like an insect while wandering in the fields' (*Heritage*, p. 389). He would have understood Thoreau's journal entry for 27 July 1840, 'Nature will bear the closest inspection. She invites us to lay our eye level with her smallest leaf and take an insect view of its plain.' Such a view is instinctive to a naturalist, but when the father of all English naturalists, Gilbert White, turns to verse, his viewpoint remains conventional, elevated:

> Romantic spot! from whence in prospect lies
> Whate'er of landscape charms our feasting eyes
> . . . .
> Now climb the steep, drop now your eye below,
> Where round the blooming village orchards grow;
> There, like a picture, lies my lowly seat,
> A rural, shelter'd, unobserv'd retreat.[38]

The 'insect view' not only undermines the neo-classical calm, but it is also closely allied to natural history, to that scientific hunger for empirical data which was to have such momentous results in the nineteenth century. Gray's annotated copy of Linnaeus, Crabbe's botanical studies, Gilbert White's *Journals*, Coleridge's Note-books, Clare's Natural History Letters, are pioneering works in the closer study of nature which preceded Darwin. As Swift knew when he wrote *Gulliver's Travels*, human beings look a lot less beautiful and important, although vastly more dangerous, when seen from an 'insect view'. But the literary expression of such a viewpoint is much rarer than one might suppose.

This is clarified when one thinks of Clare's lowly predecessors, who without exception were culturally assimilated, and adopted linguistic and spiritual elevation in an attempt to raise themselves above their station. Perhaps the saddest example is Stephen Duck (1705-56), whose poem 'The Thresher's Labour' contains many realistic descriptions of rustic labouring conditions. After he had been 'discovered' by Queen Caroline, presented with a benefice and the custodianship of her strange folly in Richmond Park, Merlin's Cave, Duck's talent dwindled to a coy mediocrity, as when he revisits the scenes of his past labours:

> Straight Emulation glows in ev'ry Vein;
> I long to try the curvous Blade again
>
> . . . .
>
> Behind 'em close, I rush the sweeping steel;
> The vanquish'd Mowers soon confess my Skill.[39]

Robert Dodsley, The Muse in Livery, wrote in 1731 'An Epistle from a Footman in London To the Celebrated Stephen Duck'. He prophesies smiling prospects for them both:

> So you and I, just naked from the Shell,
> In chirping Notes our Future singing tell;
> Unfeather'd yet, in Judgment, Thought, or Skill,
> Hop round the Basis of Parnassus' Hill.[40]

The Gods upon Parnassus eventually encouraged the fledgling to higher things, and Dodsley became a successful publisher. Other poets of lowly birth who wrote topographical verse in the conventional mould were James Woodhouse, Robert Tatersal, Ann Yearsley, Henry Jones, and Robert Bloomfield.[41]

In his early years as a writer, Clare talks of his own life and hopes, using the conventional and outworn metaphors, a terminology derived from his wide reading in eighteenth-century poetry. (Among poets in the landscape tradition, Clare was deeply read in Milton, Gray, Collins, Thomson, Goldsmith, Cowper, and Wordsworth.) The acquisition of his mature style, which largely dispenses with trite metaphor and which employs the paraphernalia of topographical verse in a fresh dimension, is a hard-earned process. In 'Helpstone', written in 1809, when he was 16, Clare uses the analogy of birds which, like Robert Dodsley's, 'Hop round the Basis of Parnassus' Hill':

> So little birds in winters frost and snow
> Doom'd (like to me) wants keener frost to know
> Searching for food and 'better life' in vain
> (Each hopeful track the yielding snows retain)

> First on the ground each fairy dream pursue
> Tho sought in vain—yet bent on higher view
> Still chirp and hope and wipe each glossy bill
>   (PMS, A3-25)

Towards the end of another early poem, which describes a walk around Burghley Park (landscaped extensively by Capability Brown from 1754-83), Clare climbs Barnack Hill, and gives a prospect, followed, as in Dyer and Bowles, by an association with time and hope:

> There uncontroul'd I knew no bounds
>   But look'd oer cottages a crowd
> On trees and vills to farthest rounds
>   Till spires seem'd rising to a cloud
>   (PMS, A6-38)
> While tir'd with such farstretching views
>   I left the green hills sideling slope
> But O! so tempting was the muse
>   She made me wish; she made me hope
> I wish'd and hop'd that future days
>   (For scenes prophetic fill'd my breast)
> Whould grant to me a Crown of bays
>   By singing maids and shepherds drest
>   (NMS, 1-22)

Directly in the topographical tradition is 'Elegy on the Ruins of Pickworth, Rutlandshire. Hastily composed, and written with a Pencil on the Spot'.[42] Clare must have been aware of how he was almost literally making a sketch, in the manner of hundreds of amateur tourists and artists, but his reading to this date (1818) had probably not included William Combe's Doctor Syntax series (1809) with its endless ridicule of such random sketching. The 'Elegy on the Ruins of Pickworth' is unusual in that Clare very rarely strays outside his own county; his mature verse is marked by its dogged rootedness, or later, a lament over forcible uprootedness. This poem seems all the more in the topographical mode by implying that the author is in transit, that this ruin is just another sought out for comparison by a picturesque tourist, another Dyer with his sketch-book, another Gray with his

Claude glass, another Gilpin with his *camera obscura*, or another Bloomfield with his undiscriminating eye. The Pickworth Elegy rings false because Clare is consciously taking up a station, whereas his mature poetry is written by what Gilbert White called a 'stationary' man. In fact, Clare had the opportunity to describe the more varied Rutlandshire landscape intimately, for he obtained a job as a lime-burner for several months near the site of the deserted village of Pickworth. But at this stage his style is heavily imbued with influences, notably Goldsmith and Gray; indeed, his pen more than once returns an echo to the other more famous Elegy:

> A time was once—tho now the nettle grows
> In triumph oer each heap that swells the ground
> When they in buildings pil'd a village rose
> With here a Cot & there a Garden crownd
>
> . . . .
> The ale house here might stand—each hamlets boast
> & here where elders rich from ruin grows
> The tempting sign—but what was once is lost
> Who would be proud of what this world bestows?
>
> . . . .
> —Since first these ruins fell—how chang'd the scene
> What busy bustling mortals now unknown
> Have com'd & gone as tho there nought had been
> Since first oblivion calld the spot her own
>
> (PMS, A3-87)

Clare is never at home with this verse-form; if the quatrain suits Gray's 'divine truisms', it is quite inappropriate to the mature Clare's purpose, which is to catch the animation and detail of nature without imprisoning it in the frame of conventional form or manner. Clare is also uneasy with the visual demands of the perspective, with space used pictorially (the assumption of foreground, middle-distance and background), the time-projections (the retrospective use of a prospect), and the moral vision controlled by the optical vision. He repeats in a laboured way the eighteenth-century device of pointing out landmarks within an ordered design, 'With *here* a Cot & *there* a Garden crownd .../ The ale house *here* might stand—each hamlets boast/ & *here* ...'

[my italics]. This device presupposes that the landscape is seen as
a framed picture, and is analogous to the device in painting by
means of which a pointing figure draws attention to the focus of
the eye.[43]

'The Village Minstrel', the title poem of the 1821 volume, uses
'prospect' in both of its principal senses:

> Thus lubins early days did rugged roll
> And mixt in timley toil—but een as now
> Ambitions prospects fird his little soul
> And fancy soard and sung bove povertys controul
>
> (NMS, 3-167)

So run the plough-boy's fanciful dreams. Something more imaginative begins to happen later in the poem. Clare has by no
means found a voice of his own, but he is beginning to adapt the
genre to the harsh facts and often joyous emotions of his life. As
time goes on, he realizes that to talk of 'ambitions prospects' is
not only trite poetically, but hopelessly unrealistic in terms of his
own career. As a poet and as a man, he is more at home on a
molehill than on a hill:

> Upon a molehill oft he dropt him down
> To take a prospect of the circling scene
> Marking how much the village roofs thatch brown
> Did add its beauty to the budding green
> Of sheltering trees it humbly pe[e]pt between
> The waggon rumbling oer the stoney ground
> The windmills sweeping sails at distance seen
> And every scene that crowds the circling round
> Where the skye stooping seems to kiss the meeting ground
>
> (NMS, 3-169)

This verse can be taken as typical of Clare's early work: within it
one can see the pull between convention and his own voice taking place. Although he is only on a molehill, he starts out 'to take
a prospect': his problem begins in the very next phrase, 'the
circling scene'. His instinctive kaleidoscopic vision runs counter
to the demands of the prospect; in order to take a prospect, the
viewer must be static and he must look in one direction at a time
(he may, of course, like the surveyor, make several sightings from

the same spot, but in succession). Clare's overriding ambition
seems to be to catch nature's events, pictures, and sequences
simultaneously (one remembers the sun making its 'instantaneous sketches'). Furthermore, Clare does not just use the
word 'circular'; he uses the active word 'circling'. Nature seems
to be in constant motion, flitting past the poet's eye in such
kinetic profusion that it threatens to break up the comfortable
pictorial framework and the correspondingly strict verse-forms.
Clare makes this clearer by repeating the word 'circling' in con-
junction with the noun 'round' and hinting at frustration in the
verb 'crowds': 'And every scene that crowds the circling round ...'.
It is all too much for the eye to take in and rather than solve it in
the easy way by succumbing to the convention (as W. L. Bowles
was still doing as late as 1828), Clare senses that a new form of
perception will have to be invented.[44]

Clare is consciously committed to the low viewpoint, the 'insect
view', which is part of his problem:

> Oer brook banks stretching on the pasture ground
> He gazd far distant from the jocund crew
> Twas but their feats that claimd a slight regard
> Twas his his pastimes lonly to pursue
> Wild blossoms creeping in the grass to view
> Scarce peeping up the tiney bent* as high
> Betingd wi glossy yellow red or blue
> Unnamd un[n]oticd but by lubins eye
> That like low genius sprang to bloom their day & dye
>
> (NMS, 3-170)

His scientific and his poetic instincts ('Unnamd, un[n]oticd')
commit him to 'creeping in the grass'. This position reduces the
importance of the purely visual and increases the power of the
other senses. The visual, none the less, may thereby attain a sort
of hallucinatory quality, especially in Clare's later poems, where
the 'insect view' in the grasses resembles the disturbing vision of
'The Fairy Feller's Master Stroke' by Richard Dadd.

In 'A Sunday with Shepherds and Herdboys', Clare begins to
describe a prospect which would fit almost unnoticed into dozens
of eighteenth-century poems, in the tradition of 'L'Allegro', or
Dyer's 'Grongar Hill', or Thomas Warton's 'The First of April':

> And oft they sit on rising ground
> To view the landscap spreading round
> Swimming from the following eye
> In greens and stems of every dye
> Oer wood and vale and fens smooth lap
> Like a richly colourd map
>
> (*SPP* 94)

But the multiplicity of things, that aspect of life which Louis MacNeice calls 'incorrigibly plural', crowds in on all Clare's senses, and he becomes sensuously involved with his surroundings, rather than seeing nature at a distance:

> Square platts* of clover red and white
> Scented wi summers warm delight
> And sinkfoil* of a fresher stain
> And different greens of varied grain
> Wheat spindles* bursted into ear
> And browning faintly—grasses sere
> In swathy* seed pods dryd by heat
> Rustling when brushd by passing feet
>
> (Ibid.)

'Hollywell' contains another conventional prospect, but Clare then tacks about and rejects the prospect as 'fictions', in favour of the immediacy of 'nearer objects':

> & as she [fancy] turns to look again
> On nearer objects wood & plain
> So lovley truth to fictions seem
> One warms as wak'ning from a dream
> The covert hedge from either side
> The black bird flutterd terryfied
> Mistaking me for pilfering boy
> That but too oft their nests destroy
> & "prink prink prink" they took to wing
> In snugger shades to build & sing
>
> . . . .
> I opt each gate wi idle swing
> & stood to listen ploughmen sing
> While cracking whip & gingling gears
> Recalld the toils of boyis[h] years

When like to them I took my rounds
Oer elting* moulds of fallow grounds—
Wi feet neer* shooless paddling thro
The bitterest blasts that ever blue
(PMS, B2-127/127a)

Clare here brings present and past simultaneously alive with
'cracking whip & gingling gears'. Duck would rather not remember his 'shooless' feet, and would have thought the following
lines, so typical of Clare's mature work, beneath his dignity:

& neath the hanging bushes creep
For vi'let bud & primrose peep
& sigh wi anxious eager dream
For water blobs* amid the stream
& up the hill side turn anon
To pick the daiseys one by one
(PMS, B2-128)

'The Woodman' contains another rejection of the conventional
view:

The pleasing prospect does his heart much good
Tho tis not his such beautys to admire
He hastes to fill his bags wi billet wood
Well pleasd from the chill prospect to retire
To seek his corner chair and warm snug cottage fire
(PMS, B2-238a/239)

The woodman turns from a prospect which is 'pleasing' in terms
of fancy, but 'chill' in terms of his own life. There is a gruff,
countryman's realism about the line, 'Tho tis not his such beautys
to admire', and he seeks out the small but vivid comforts of his
own experience. If the aristocrat tends to live in the past, and the
bourgeois in the future, the countryman lives in the present. The
pint of ale in his hand, or his 'snug cottage fire', is of more interest to him than his ancestors or his prospects.

This uncalculating involvement with the present is well described by W. K. Richmond in his discussion of the poem 'Pleasures of Spring': 'It is written as a labourer might hoe a field of
turnips, with no eye on the ending, no thought of what is to
come next, but with a massive, unquestioning patience which

sustains the work and makes it not ignoble.' (*Heritage*, p. 400.) Clare's prose also has 'no eye on the ending, no thought of what is to come next'. In the following passage from 'The Autobiography', in spite of the word 'survey' and Clare's position on a 'mossy eminence', his eye does not pan the view as on a tripod; it zigzags from object to object like a gadfly, it has no resting place or focal point around which to frame a design, and the writing catches the feeling of breathless excitement:

I dropt down on the thymy molehill or mossy eminence to survey the summer landscape as full of rapture as now I markd the varied colors in flat spreading fields checkerd with closes of different tinted grain like the colors in a map the copper tinted colors of clover in blossom the sun-tannd green of the ripening hay the lighter hues of wheat & barley intermixd with the sunny glare of the yellow carlock & the sunset imitation of the scarlet headaches with the blue cornbottles crowding their splendid colors in large sheets over the land & troubling the cornfields with destroying beauty the different greens of the woodland trees the dark oak the paler ash the mellow lime the white poplar peeping above the rest like leafy steeples the grey willow shining chilly in the sun as if the morning mist still lingered on its cool green (*Prose* 25)

That 'survey' would not be of much use to a cartographer. The absence of spacing, design, or perspective is paralleled by the absence of punctuation; one is not given a single comma to draw one's breath, or to feel that this is here, and that is there, or that this impression comes after the previous one. Instead one is given a kaleidoscopic vision which shares the virtues of Clare's mature poetry—freshness, rhythmical subtlety, euphony, and the crystallization of emotion into haunting images.

The same qualities can be found in the sonnets of Clare's maturity (from *c*.1821 onwards). If the sonnet is a 'moment's monument', it is not difficult to understand Clare's success with this form, given his fascination with the ephemeral, the local, and the microscopic. In his use of this form, Clare is often at the furthest remove from neo-classical tenets, unconcerned as he is with punctuation and perspective (in the topographical poem the two are connected), and the use of space and time as moral emblems. He extends the range of the sonnet, often in highly irregular ways—seven couplets strung together, for example. But Clare's mature sonnets are never stilted, and they impose upon him the pressure of selection from the endless detail at his disposal. They share Thomas Bewick's intense perception of the minutiae of nature and have Bewick's compressed harmony. They are tail-pieces (or tale-pieces) which catch the texture and multiplicity of the world without imprisoning its bright details:

> Now sallow* catkins once all downy white
> Turn like the sunshine into golden light
> The rocking clown* leans oer the spinny rail
> In admiration at the sunny sight
> The while the blackcap doth his ears assail
> With a rich and such an early song
> He stops his own and thinks the nightingale
> Hath of her monthly reckoning counted wrong
> 'Sweet jug jug jug' comes loud upon his ear
> Those sounds that unto may by right belong
> Yet on the awthorn scarce a leaf appears
> How can it be—spell struck the wandering boy
> Listens again—again the sound he hears
> And mocks it in his song for very joy
>
> (*SPP* 72)

There is no projection into past and future, no use of space in three dimensions, and no moralizing. Instead, it is the sound of the blackcap which spaces the poem in an arbitrary way, which the eighteenth-century eye might have found disorientating; to the 'rocking clown' it is freedom.

Clare writes in 'The Autobiography', 'I usd to drop down under a bush & scribble the fresh thoughts on the crown of my hat' (*Prose*, p. 52). This passage gives fuel to those who would like to think of the Romantic poet as effortlessly responding to nature as if he were an Aeolian harp in the breeze. The spontaneity of Clare's mature work, verse and prose, is not as simple as that. If he can achieve an accurate sketch from nature, he had laid the foundations of his art in the studio, and he had copied the masters. Two of Clare's masters were Thomson and Cowper. Thomson's strong sense of movement influenced Clare, and Cowper's delicate descriptions scattered throughout *The Task* also had much appeal for him. If one abstracts one of Cowper's

vignettes from its narrative context, the effect is not unlike a
Clare sonnet:

> Forth goes the woodman, leaving unconcern'd
> The cheerful haunts of man; to wield the axe
> And drive the wedge, in yonder forest drear,
> From morn to eve his solitary task.
> Shaggy, and lean, and shrewd, with pointed ears
> And tail cropp'd short, half lurcher and half cur—
> His dog attends him. Close behind his heel
> Now creeps he slow; and now, with many a frisk
> Wide-scamp'ring, snatches up the drifted snow
> With iv'ry teeth, or ploughs it with his snout;
> Then shakes his powder'd coat, and barks for joy.
> Heedless of all his pranks, the sturdy churl
> Moves right toward the mark; nor stops for aught,
> But now and then with pressure of his thumb
> T'adjust the fragrant charge of a short tube
> That fumes beneath his nose: the trailing cloud
> Streams far behind him, scenting all the air.[45]

This is an unframed landscape, as in a Bewick woodcut, full of
incident, movement and immediacy. But of course it is only a
brief interlude from 'The Winter Morning Walk', which is main-
ly concerned with Monarchy, Liberty, Patriotism, and Deism,
and which ends with an 'Address to the Creator'.

Clare's sonnet 'Winter Fields' describes a similar scene:

> O FOR a pleasant book to cheat the sway
> Of winter—where rich mirth with hearty laugh
> Listens and rubs his legs on corner seat
> For fields are mire and sludge—and badly off
> Are those who on their pudgy* paths delay
> There striding shepherd seeking driest way
> Fearing nights wetshod feet and hacking cough
> That keeps him waken till the peep of day
> Goes shouldering onward and with ready hook
> Progs* oft to ford the sloughs that nearly meet
> Accross the lands—croodling* and thin to view
> His loath dog follows—stops and quakes and looks
> For better roads—till whistled to pursue
> Then on with frequent jump he hirkles* through
>
> (SPP 139)

But how unmistakably Clare's own these lines are, not just because of the dialect. Like a drop of water, this sonnet reflects a whole world. The effect is achieved by deft strokes of description but goes far beyond the visual or pictorial connotations of that word. The poem conveys the boredom of winter, the hardships of those living close to the soil, and the texture of weather which is both impersonal event and emotional reaction; the personification 'rich mirth' does not jar with the earthy phrase 'rubs his legs'—all of this slips into the reader's mind effortlessly, so charmed is he by the evocation of the dog manoeuvring his way round the puddles.

In order to experience the 'instantaneous sketches' of landscape rather than a static view in a single frame, Clare lowers his eye level until he substitutes a molehill for Parnassus. The observation of the naturalist blends, as always in Clare, with the emotion of the poet:

> —Five eggs pen-scribbled over lilac shells
> Resembling writing scrawls which fancy reads
> As natures poesy & pastoral spells
> They are the yellowhammers & she dwells
> A poet-like—where brooks & flowery weeds
> As sweet as Castaly to fancy seems
> & that old molehill like as parnass hill
> On which her partner haply sits & dreams
> Oer all his joy of song—so leave it still
> A happy home of sunshine flowers & streams
>   (*MC* 239)

The yellow-hammer is 'poet-like' and 'her partner' is not only the male bird but Clare himself, versed in 'natures poesy'. This poem is an echo of the earlier, more complex 'Shadows of Taste', in which the yellow-hammer is also praised as a poetic model worthy of imitation:

> Taste with as many hues doth hearts engage
> As leaves and flowers do upon natures page
> Not mind alone the instinctive mood declares
> But birds and flowers and insects are its heirs
> Taste is their joyous heritage and they

All choose for joy in a peculiar way
Birds own it in the various spots they chuse
Some live content in low grass gemmed with dews
The yellow hammer like a tasteful guest
Neath picturesque green molehills makes a nest
Where oft the shepherd with unlearned ken
Finds strange eggs scribbled as with ink and pen
He looks with wonder on the learned marks
And calls them in his memory writing larks
(SPP 112)

The molehills have not only become Parnassus, they are described as 'picturesque' as well. This adjective crops up frequently
in Clare's mature poetry, but it is rarely used in a derivative way;
it is used to create what an art historian has called a 'micro-
panorama':[46]

Rude architect rich instincts natural taste
Is thine by heritage—thy little mounds
Bedecking furze clad heath & rushy waste
Betraced with sheep tracks shine like pleasure grounds
No rude inellegance thy work confounds
But scenes of picturesque & beautiful
Lye mid thy little hills of cushioned thyme
On which the cowboy when his hands are full
Of wild flowers leans upon his arm at rest
As though his seat were feathers—when I climb
Thy little fragrant mounds I feel thy guest
& hail neglect thy patron who contrives
Waste spots for the[e] on natures quiet breast
& taste loves best where thy still labour thrives
(PMS, A54-400)

This sonnet is a deliberate inversion of the assumptions and
devices of the topographical tradition. The wayward sheep-tracks
on the heath 'shine like pleasure grounds'; the poet mimics the
tradition by pretending to 'climb' the hillocks, and he describes
himself as the mole's 'guest', an admission the topographical
poet could not have made, dedicated as he was to creating the
illusion that he was monarch of all he surveyed. Both mole and
poet now create their own space, and are not passively contained,

as objects, by an intellectually and aesthetically imposed order; they both 'choose for joy in a peculiar way'.

The ripening of Clare's vision evidenced by these quotations should be sufficient to refute John Middleton Murry's claim that Clare 'had nothing of the principle of inward growth which gives to Wordsworth's most careless work a place within the unity of a great scheme'.[47] In spite of Eric Robinson's and Geoffrey Summerfield's refutation of this same passage (*SPP*, p. xvii), the conscious artifice of Clare's best work is still undervalued. This is the price Clare paid for wanting to be too like nature—it is a very Romantic dilemma.

Just as Denham had assumed that, given the presence of a poet, Cooper's Hill could be made Parnassus, so Clare assumes that the meanest details of nature are poetic material. The mole-hills need no reference beyond themselves; they simply have to be loved, and the poet's 'natural taste', in tune with 'natures poesy', will do the rest:

> So where the Muses & their train resort,
> *Parnassus* stands; if I can be to thee
> A Poet, thou *Parnassus* art to me.

If Denham uses his eye like a telescope, Clare, having emancipated himself from the Denham tradition, turns out to be quite unlike those Romantics whose approach to nature might at first seem similar; he feels almost as uneasy with the constant need for elevation of sentiment in Wordsworth and Keats as he does with the ambiguities of the topographical tradition. Clare's reaction to Keats is that 'behind every rose bush he looks for a Venus & under every laurel a thrumming Appollo' (*Prose*, p. 223). The flowers at Clare's feet took all his love, and not those things that they symbolized; he happily ignored Lamb's advice to 'Transplant Arcadia to Helpstone.' (*Heritage*, p. 175.) He also chose to ignore Keats's comment, relayed to him in a letter of 1820 by John Taylor, 'that the Description too much prevailed over the Sentiment.' (Ibid., p. 120.) Sentiment was overemployed by the Victorians until it became sentimentality. Clare's kinetic and microscopic descriptions, devoid of ethical, patriotic, or religious comment, link him rather to twentieth-century sensibilities than to nineteenth. Yet he is rooted in his landscape in a way no modern man can ever be again. For a short period of magic, Clare is able to fuse objective naturalistic detail with subjective poetic response. The loss of this balance lay behind the intense nostalgia of so much Victorian art and poetry; indeed, the rapid divorce of 'the two cultures' in Clare's manhood must have contributed to his need for an asylum. The next chapter is an examination of Clare's use of natural history, a process partly encouraged by the development of the Picturesque, and one markedly foreign to neo-classical ideals.

2

'Thy Wild Seclusions'

For the neo-classical artist, the particular is only admissible if it can be merged in the general and the ideal; if it is too eccentric, too vivid (for much of the eighteenth century, the adjective 'picturesque' is used synonymously with vivid or striking), the effect of balance and decorum is lost. Pope writes in the postscript to his translation of the *Odyssey*:

> The question is, how far a Poet, in pursuing the description or image of an action, can attach himself to *little circumstances*, without vulgarity or trifling? what particulars are proper, and enliven the image; or what are impertinent, and clog it? (*Pope: Poems* x. 387)

The development of picturesque vision, comic and faddish as much of the theory appears in retrospect, steadily undermined such a viewpoint. Picturesque artists sought rough surfaces, variety, and shaggy outlines (if a horse was beautiful, then a goat was picturesque). As the century progressed, words such as 'proper' and 'vulgar' became irrelevant to many visual experiences; they were replaced by a more morally neutral vocabulary which dealt with levels of interest and sensation rather than propriety. As poets increasingly incorporated the details of natural history into their work, they inevitably employed specific rather than generic description. The more Clare explored the hidden corners of unimproved nature, the more he found his true voice.

As early as 1777, John Aikin in his *Essay on the Application of Natural History to Poetry* had complained of the lack of original observation in poetry. Aikin, like Clare, did not avoid picturesque detail but assimilated it in order to make it serve his ideal of accuracy. Where Gilpin saw the Picturesque as largely avoiding detail (and his sketches bear this out), Aikin believed that the picturesque eye is allied to the scientific eye. Although Clare had probably not read Aikin's work, it is important to take it into account, for Aikin saw that the Picturesque and natural history were very far from mutually exclusive:

As the artist who has not studied the body with anatomical precision, and examined the proportions of every limb, both with respect to its own several parts, and the whole system, cannot produce a just and harmonious representation of the human frame; so the descriptive poet, who does not habituate himself to view the several objects of nature minutely, and in comparison with each other, must ever fail in giving his pictures the congruity and animation of real life.[1]

Aikin quotes four examples of the description of the fall of evening and detects 'the hand of a copyist' in the passages, rather than 'the strokes of an observer' (pp. 6-8). The four passages are:

> ... to black Hecat's summons
> The shard-born beetle with his drowsy hums
> Hath rung night's yawning peal.
> (Shakespeare)

> ... both together heard
> What time the gray-fly winds her sultry horn,
> Battening our flocks with the fresh dews of night.
> (Milton)

> Save where the beetle wheels his droning flight.
> (Gray)

> Or where the beetle winds
> His small, but sullen horn,
> As oft he rises midst the twilight path,
> Against the pilgrim borne in heedless hum.
> (Collins)

One can add Clare's beetle passage in 'Summer Evening' to judge whether it is a fresh observation:

> From the hedge the beetles boom
> Heedless buz and drousy hum
> Haunting every bushy place
> Flopping in the labourers face
> (NMS, 1-1)

All these passages have a certain beauty, but there is no doubt that Clare has been more successful (from Aikin's point of view and the modern reader's) in catching the 'animation of real life'. One must not assume, of course, that the first four authors intended to supply more than an evocation, in order to set up a series of almost endless echoes in the mind of the reader, and all of them would have assumed their readers to be well read, especially in the classics. One can see why Clare admired Collins as one of those poets who 'went to nature for their images' (*Prose*, p. 175), but Collins is closer to the first three than he is to Clare.

Clare's 'drousy hum' is a direct echo of Shakespeare's 'drowsy hums', but there the similarity ends. The monosyllabic directness of the word 'boom' does the work of whole phrases such as 'winds her sultry horn' (Milton), 'wheels his droning flight' (Gray), and 'winds/ His small, but sullen horn' (Collins). I am not suggesting that nothing is lost, but on the other hand, something has been gained. One could play a literary game with these passages, in order to point up the difference in tone:

> ... both together heard
> What time the gray-fly booms,
> Battening our flocks with the fresh dews of night.
> (Milton?)

> Or where the beetle winds
> His small, but sullen horn,
> As oft he rises midst the twilight path,
> Against the labourer in heedless hum.
> (Collins?)

Finally, whereas the first four passages admit such a minute specimen as the beetle, it is really the *sound* of the beetle that is admitted, as a convention. The other writers are listening to a literary echo, but Clare is listening to and observing *this* insect which flies out of *this* hedge and haunts 'every bushy place'. Although Clare's passage is, like the others, auditory, it differs from them in giving the beetle an autonomous existence, and a large part of this existence is visual for the fascinated naturalist such as Clare. Aikin uses the word 'picturesque' to describe the minutiae of nature, such as birds, and not just buildings, bridges,

trees, or views. Uvedale Price writes in relation to buildings,
'Whatever then has strong attractions as a visible object, must
have a character; and that which has strong attractions for the
painter, and yet is neither grand nor beautiful, is justly called picturesque.'[2] Aikin and Clare extend this criterion to beetles.

In his recommendation of the use of natural history in poetry,
and his choice of Virgil's *Georgics* and Thomson's *Seasons* as
models, Aikin's *Essay* was timely. But although it appeared
revolutionary when compared with the stately neo-classical
judgements of Reynolds and Johnson, the trend towards greater
detail and accuracy had been going on for some time. Even Aikin
drew the line judiciously short of what we would now call
'photographic realism'. He writes of Cowper's poems in *Letters
to a Young Lady* (1804), 'They would resemble the Dutch style of
painting, did not the writer's elegance of taste generally lead him
to select only such objects as are capable of pleasing or picturesque effect.' (p. 291.) One wonders if Cowper's cucumber
georgic was 'capable of pleasing or picturesque effect':

> The stable yields a stercoraceous heap,
> Impregnated with quick fermenting salts,
> And potent to resist the freezing blast:
> For, ere the beech and elm have cast their leaf
> Deciduous, when now November dark
> Checks vegetation in the torpid plant
> Expos'd to his cold breath, the task begins.
> (*Cowper: PW* 174)

Wordsworth wrote in a letter to Samuel Rogers (29 September
1808), 'nineteen out of twenty of Crabbe's Pictures are mere
matters of fact; with which the Muses have just about as much to
do as they have with a collection of medical reports, or of law
cases.'[3] These two criticisms assume that too much detail is inar-
tistic, that art always involves a distancing and idealization, and
that 'mere matters of fact' are always inimical to poetry. It is an
assumption that Pope, Dyer, and Gray would have taken for
granted. Pope tends to smooth out the over-vivid detail in his
Homer translation (a practice Aikin condemns), and Gray is of
especial interest in that his poetry compresses or omits the abun-

(*Left*) Sketch of a moth on the inside back cover of one of Clare's manuscript notebooks at Northampton

(*Below*) Part of a page from Thomas Gray's annotated copy of Linnaeus' *Systema Naturae*, showing Gray's drawings of moths

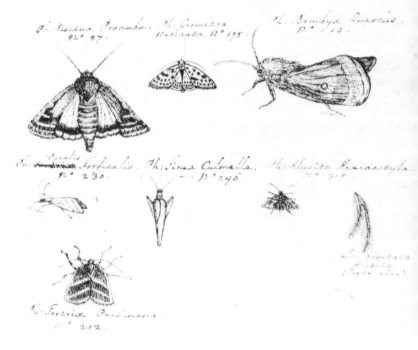

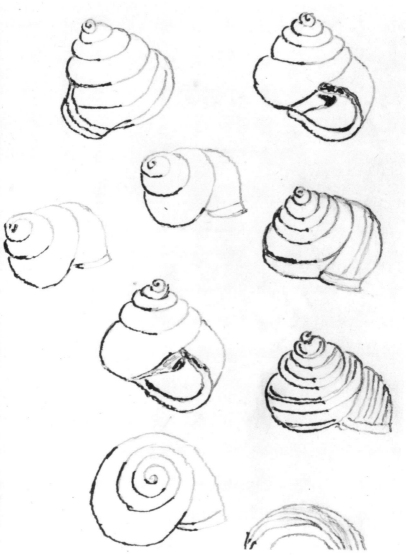

Drawing of shells from Clare's manuscript notebooks

dant detail of his knowledge. We have seen Gray's poetic treatment of a beetle, and a reader might assume that he had never
looked at nature precisely; but in the words of E. D. H. Johnson:

> Gray immersed himself in the study of natural history during the last
> decade of his life. Like Gilbert White, he kept annual calendars in
> which he meticulously recorded temperature readings and weather con-
> ditions, and made detailed notations on such matters as the budding of
> foliage and the springing of crops, the first arrival of blossoms and
> fruits, the migration of birds with their songs and nesting habits....
> Many of the notes are illustrated by sensitive pen-and-ink drawings of
> birds and insects.[4]

Johnson reproduces a facsimile of a page from Gray's annotated
copy of Linnaeus's *Systema Naturae*, which is littered with drawings of moths, detailed down to the minutest markings on their
speckled wings. Clare's manuscript notebooks at Peterborough
and Northampton also contain drawings of birds, steeples, and
shells. On the inside back cover of one notebook there is a coloured sketch of a moth as detailed and as sensitive as Gray's (see
previous pages). Gray's moths stay pinned down, as it were,
meticulously classified and their names Latinized, whereas Clare
attempts to give his insects and birds the freedom of his verse.

Linnaeus's work was a major factor in the emergence of a new
approach to landscape, in which ecological relationships were
more important than relationships created by neo-classical
tenets. In his essay 'Discovery and Exploration: The European
and the Pacific', Bernard Smith writes of the Royal Society,
which had been founded in 1660:

> ... it was the empirical approach of the Society and not the neo-classical
> approach of the Academy [Royal Academy] which flourished under the
> impact of the new knowledge won from the Pacific. For though the
> discovery of the Society Islands gave initial support to the belief that a
> kind of tropical Arcadia inhabited by men like Greek gods existed in the
> South Seas, increasing knowledge not only destroyed the illusion but
> also became a most enduring challenge to the supremacy of neo-classical
> values in art and thought.[5]

Bernard Smith discusses how the study of nature in detail involved the discovery of a large number of independent localities, each

with its ecological system. Clare's Helpston could be taken as one of these independent localities, and his work as a botanist and an ornithologist was a humble parallel to the Pacific explorations in that it reveals the richness, the interrelated details, and the uniqueness of his locale. It is not for nothing that E. P. Hood in *The Literature of Labour* (1851) called Clare 'an English Audubon' (*Heritage*, p. 262).

In reading the following extract from Bernard Smith's essay, we can bring to mind Thomas Bewick's detailed portraits of bird and animal life, in which the environment is so accurately related to the habits and character of each bird or animal. It is surprising that Clare rarely mentions Bewick's work,[6] but his library contains some finely illustrated collections, including *A Natural History of all the most remarkable Quadrupeds, Birds, etc.* (1820) by J. Macloc; *The Florist's Directory* (1822) by James Maddock; *The Natural History of Quadrupeds, and Cetaceous Animals* (1811); and *The Natural History of Birds* (1834) by Robert Mudie. Bernard Smith writes:

An early example of the type of relationship which art began to take over from science at this time is to be found in botanical and zoological illustration. One of the important features of the description of a new species was the description of its habitat. Consequently, illustrators of animals usually placed them in their appropriate environmental setting. During the period under discussion this became a more frequent practice in the illustration of plants also. The relationship which existed between a species and its habitat could, however, be extended to all the species peculiar to a particular habitat. The placing of plants, animals, and primitive peoples in their appropriate environmental situation became a matter of increasing importance for the landscape-painter. (*Backgrounds to British Romantic Literature*, 24)

For personal as well as artistic reasons, it was 'a matter of increasing importance' for Clare to place his flora and fauna and his people in their appropriate habitat. And since it is very likely that Clare was greatly influenced by those botanical and ornithological illustrations, what better word than 'picturesque' was there to lead his attention into the secret places, where a great deal was to be seen which Gilpin, Price, and Payne Knight had passed by? Price talks of Vanbrugh as a picturesque architect (*Essays* II. 212-24); in the following extract, Clare describes picturesque architects of more importance to him. The words 'gardens', 'taste', 'architects', 'picturesque', and 'pictures' are used in a way which subverts their usual context, and changes much of their meaning:

> Then een the fallow fields appear so fair
> The very weeds make sweetest gardens there
> And summer there puts garments on so gay
> I hate the plough that comes to dissaray
> Her holiday delights—and labours toil
> Seems vulgar curses on the sunny soil
> And man the only object that distrains
> Earths garden into deserts for his gains
> Leave him his schemes of gain—tis wealth to me
> Wild heaths to trace—and note their broken tree
> Which lightening shivered—and which nature tries
> To keep alive for poesy to prize
> Upon whose mossy roots my leisure sits
> To hear the birds pipe oer their amorous fits
> Though less beloved for singing then the taste
> They have to choose such homes upon the waste
> Rich architects—and then the spots to see
> How picturesque their dwelling makes them be
> The wild romances of the poets mind
> No sweeter pictures for their tales can find
> And so I glad my heart and rove along
> Now finding nests—then listening to a song
> Then drinking fragrance whose perfuming cheats
> Tinges lifes sours and bitters into sweets
>
> (*SPP* 80)

Clare knows what he is doing; he is assimilating these words to *his* way of thinking and seeing, not the other way around. The words no longer look out of place or sound vague; he has made them his own. It is evident that he knew enough about the theory and practice of the Picturesque to be able to make up his own mind, and he decided that much of it was not for him. But he could employ its technique of training the eye for his own ends, and at his best, he can be confused with no other poet.

The accuracy of Clare's bird descriptions is by no means un-precedented. What does mark Clare off from most other eighteenth-century poets is his inability to be didactic in the man-ner of the true georgic. His best descriptions come to us un-embellished. In spite of Aikin's complaint about the paucity of accurate description, the list of poets who extended the reader's knowledge and appreciation of the external world is almost endless. D. L. Durling gives the following: 'Gay, Browne, Mallet, Ralph, Savage, Bruce, Dyer, Rolt, Akenside, John Scott, John Wilson, Stevenson, Falconer, Crowe, Mason, Cowper, Leyden, Hurdis, Budworth, the Gisbornes, Bidlake, Bloomfield, Bowles, Drummond, Grahame, Cririe, and Clare'.[7]

Hurdis, Thomas Gisborne, and James Grahame are particular-ly good on birds. Here is Hurdis:

> No longer now assembles as of late,
> Gregarious only in the winter hour,
> Bird of the sky baptiz'd, the speckled lark.
> Oft o'er the plain inert or fallow then
> In flight circuitous the nimble flock
> Swam eddying, or with sudden wheel revers'd
> Show'd their transparent pinions to the sun.
>
> (Ibid. 170)

Thomas Gisborne's *Walks in a Forest* (1795) ran to eight editions by 1813. He is influenced by Thomas Pennant (to whom both Aikin's *Essay* and much of White's *Selborne* is addressed), who had advised the poets to go to nature in his *British Zoology:*

> With shrill and oft-repeated cry
> Her angular course, alternate rise and fall,
> The woodpecker prolongs; then to the trunk
> Close-clinging, with unwearied beak assails
> The hollow bark; through every cell the strokes
> Roll the dire echoes that from wintry sleep
> Awake her insect prey; the alarméd tribes
> Start from each chink that cleaves the mouldering stem ...
>
> (Ibid. 164)

James Grahame's *The Birds of Scotland* is perhaps the most at-tractive answer to Aikin's complaint before Clare. He describes the wren:

> 'Neath some old root,
> From which the sloping soil, by wintry rains,
> Has been all worn away, she fixes up
> Her curious dwelling, close, and vaulted o'er,
> And in the side a little gateway porch
>
> . . . .
>
> Nor always does a single gate suffice
> For exit, and for entrance to her dome;
> For when (as sometimes haps) within a bush
> She builds the artful fabric, then each side
> Has its own portico
>
> . . . .
>
> Fifteen white spherules, small as moorland hare-bell,
> And prettily bespecked like fox-glove flower,
> Complete her number.[8]

The formality of the architectural words 'porch', 'dome', and 'portico' does not detract here from the fresh observation. Aikin had written, 'the poet should think it incumbent upon him to discover and investigate *new facts*, as well as to frame *new combinations of words*.' (*Essay*, p. 132.) He also recommended the migration of birds as an excellent new topic for poetry, a challenge James Grahame took up in his account of swallows and the corncrake (*The Birds of Scotland*, pp. 50-1). This is a good example of how scientific knowledge quickly became assimilated by poetry—even so skilled a naturalist as Gilbert White (1720-93) had an absurd theory about swallows.

Hurdis, Gisborne, and Grahame all draw attention to details the ordinary observer might miss, and their pictures are certainly more than 'mere matters of fact'. But Clare's emotions are more vivid and haunting, his eye even more precise, and his continual sense of wonder would gain respect, even if it were less perfectly expressed. 'To the Snipe' is a poem beyond the range of most of those on Durling's list, and one regrets that Clare did not use this verse-form more often:

> Lover of swamps
> The quagmire over grown
> With hassock tufts of sedge—where fear encamps
> Around thy home alone
>
> . . . .

For here thy bill
Suited by wisdom good
Of rude unseemly length doth delve and drill
The gelid* mass for food
....
Boys thread the woods
To their remotest shades
But in these marshy flats these stagnant floods
Security pervades

....
In these thy haunts
Ive gleaned habitual love
From the vague world where pride and folly taunts
I muse and look above

....
I see the sky
Smile on the meanest spot
Giving to all that creep or walk or.flye
A calm and cordial lot

Thine teaches me
Right feelings to employ
That in the dreariest places peace will be
A dweller and a joy

(*SPP* 69-72)

The infinite variousness of birds brings out the best in Clare's writing. They never lack independence and vitality, even when caged in unnecessary punctuation:

I love to hear the evening crows go by
And see the starnels* darken down the sky;
The bleaching stack the bustling sparrow leaves,
And plops with merry note beneath the eaves.
The odd and lated pigeon bounces by,
As if a wary watching hawk was nigh,
While far and fearing nothing, high and slow,
The stranger birds to distant places go;
While short of flight the evening robin comes
To watch the maiden sweeping out the crumbs,
Nor fears the idle shout of passing boy,

But pecks about the door, and sings for joy;
Then in the hovel where the cows are fed
Finds till the morning comes a pleasant bed.
(Poems II. 363)

The verbs 'darken', 'plops', and 'bounces' catch the movements
of the respective birds with a precision born of long observation,
and here Clare out-Thomsons Thomson in his use of active,
kinetic verbs. The accuracy of Thomson's bird descriptions is
shown in such a passage as this from 'Spring':

> With stealthy wing,
> Should some rude foot their woody haunts molest,
> Amid a neighbouring bush they silent drop,
> And whirring thence, as if alarmed, deceive
> The unfeeling schoolboy. Hence, around the head
> Of wandering swain, the white-winged plover wheels
> Her sounding flight, and then directly on
> In long excursion skims the level lawn
> To tempt him from her nest. The wild-duck, hence,
> O'er the rough moss, and o'er the trackless waste
> The heath-hen flutters, pious fraud! to lead
> The hot pursuing spaniel far astray.[9]

Thomson's world is animated, but it is a controlled and 'literary'
one. Clare has no time to interpolate such phrases as 'pious
fraud!'; if he had paused to moralize, his birds would have flown
away.

Clare's descriptions of birds are never sentimental, because
they spring from precise observation. They can be compared
with the language used by Daines Barrington in a scientific paper
on the nightingale, which is quoted by Aikin in his *Essay:*

When this bird, (a very fine caged nightingale which Mr. B. kept) *sang
its song round*, in its whole compass, I have observed sixteen different
beginnings and closes, at the same time that the intermediate notes were
commonly varied in their succession with such judgment as to produce
a most pleasing variety.... Whenever respiration, however, became
necessary, it was taken with as much judgment as by an opera singer.

But it is not only in tone and variety that the nightingale excels; the
bird also sings (if I may so express myself) with superior judgment and
taste.

I have therefore commonly observed that my nightingale began softly like the ancient orators; reserving its breath to swell certain notes, which by this means had a most astonishing effect, and which eludes all verbal description. (pp. 137-9)

Clare does not feel the need for Barrington's hesitant parenthesis '(if I may so express myself)', and in another poem he attempts a 'verbal description' of the 'astonishing effect' of the nightingale's song:

> And nightingales O I have stood
> Beside the pingle * and the wood
> And oer the old oak railing hung
> To listen every note they sung
> And left boys making taws * of clay
> To muse and listen half the day
> The more I listened and the more
> Each note seemed sweeter then before
> And aye so different was the strain
> Shed scarce repeat the note again
> —'Chew-chew Chew-chew'—and higher still
> 'Cheer-cheer Cheer-cheer'—more loud and shrill
> 'Cheer-up Cheer-up cheer-up—and dropt
> Low 'Tweet tweet tweet jug jug jug' and stopt
> One moment just to drink the sound
> Her music made and then a round
> Of stranger witching notes was heard
> As if it was a stranger bird
> 'Wew-wew wew-wew chur-chur chur-chur
> 'Woo-it woo-it'—could this be her
> 'Tee-rew Tee-rew tee-rew tee-rew
> 'Chew-rit chew-rit'—and ever new
> 'Will-will will-will grig-grig grig-grig'
> The boy stopt sudden on the brig *
> To hear the 'tweet tweet tweet' so shill
> Then 'jug jug jug'—and all was still
> A minute—when a wilder strain
> Made boys and woods to pause again
>
> (SPP 122-3)

This passage may not be the highest form of poetry, but it has the patient eye and ear of the ornithologist, and would surely have pleased both Barrington and Gilbert White. It would not have pleased Sir Joshua Reynolds. In his seventh discourse, discussing taste, he writes:

As these prejudices become more narrow, more local, more transitory, this secondary taste becomes more and more fantastical; recedes from real science; is less to be approved by reason, and less followed in practice... (*Discourses on Art* 110)

Clare's passage could be described as becoming 'more narrow, more local, more transitory', but it could not be accused of receding 'from real science'. On the contrary. But Reynolds is using the word 'science' entirely differently from the way Barrington and Clare understood it. The same applies to the words 'taste' and 'nature':

The beginning, the middle, and the end of every thing that is valuable in taste, is comprised in the knowledge of what is truly nature; for whatever notions are not conformable to those of nature, or universal opinion, must be considered as more or less capricious. (Ibid 111)

Clare's particularized description of the nightingale's song, variations included—'Shed scarce repeat the note again'—would have struck Reynolds as unnatural, capricious, and therefore tasteless. For Barrington and Clare, however, the nightingale is endowed with natural taste, and the poet or natural philosopher can learn to be similarly endowed by patient and accurate attention:

I can sit at my window here and hear the nightingale singing in the orchard & I attempted to take down her notes but they are so varied that every time she starts again after the pauses seems to be something different to what she uttered before & many of her notes are sounds that cannot be written the alphabet having no letters that can syllable the sounds (PMS, A58-10)

The extreme Romantic sees everything through a mood, relates everything to himself. This will be clear if we compare Clare's poem 'Sand Martin' to a description of the sand-martin by Gilbert White. White's tone and the very shape of his syntax allow one to remain in a state of detached curiosity; feeling is certainly not absent but the emotion which powers the curiosity and the loving observation is never allowed to interfere with the balanced decorum of the proceedings:

> But it is much to be regretted that it is scarce possible for any observer to be so full and exact as he could wish in reciting the circumstances attending the life and conversation of this little bird, since it is *fera naturâ*, at least in this part of the kingdom, disclaiming all domestic attachments, and haunting wild heaths and commons where there are large lakes ... (*Selborne* 178)

Notice also how the Latin phrase helps to keep the bird at a respectable distance. The equilibrium of White's feelings is something Clare finds impossible to maintain, in spite of his accurate observations:

> Thou hermit haunter of the lonely glen
> And common wild and heath—the desolate face
> Of rude waste landscapes far away from men
> Where frequent quarrys give thee dwelling place
> With strangest taste and labour undeterred
> Drilling small holes along the quarrys side
> More like the haunts of vermin than a bird
> And seldom by the nesting boy descried
> Ive seen thee far away from all thy tribe
> Flirting* about the unfrequented sky
> And felt a feeling that I cant describe
> Of lone seclusion and a hermit joy
> To see thee circle round nor go beyond
> That lone heath and its melancholly pond
>
> (*SPP* 69)

This poem reveals Clare's poetic strength as much as it explains his extreme vulnerability as a man. There is an unobtrusive use of repetition, which produces effective echoes, such as 'frequent quarrys' and 'unfrequented sky'. But the most striking aspect of the poem is Clare's sense of identification with the sand-martin, who is a 'hermit' and 'far away from men'. Clare is aware of the strangeness and power of his emotion and almost admits artistic defeat, 'a feeling that I cant describe'. The emotional force of his involvement has driven this poem beyond description to a passionate lonely ecstasy. The poem is ecstatic because Clare is out of himself, beyond normal space and time in a dimension of instinctive joy. And it is fitting that the bird's motion is circular, its world self-contained and self-absorbed. We are at the opposite pole here from aesthetic detachment. This is not rhymed natural history, this is Clare's personal history enacted by natural events.[10]

A striking similarity of language and observation occurs between Clare's poem 'Mouse's Nest', and the following paragraph by Gilbert White:

A still more remarkable mixture of sagacity and instinct occurred to me one day as my people were pulling off the lining of a hotbed, in order to add some fresh dung. From out of the side of this bed leaped an animal with great agility that made a most grotesque figure; nor was it without great difficulty that it could be taken; when it proved to be a large white-bellied field-mouse with three or four young clinging to her teats by their mouths and feet. It was amazing that the desultory and rapid motions of this dam should not oblige her litter to quit their hold, especially when it appeared that they were so young as to be both naked and blind! (*Selborne* 156)

Clare imbues such an incident with his characteristic emotional tone:

> I found a ball of grass among the hay
> And proged* it as I passed and went away
> And when I looked I fancied somthing stirred
> And turned agen and hoped to catch the bird
> When out an old mouse bolted in the wheat
> With all her young ones hanging at her teats
> She looked so odd and so grotesque to me
> I ran and wondered what the thing could be
> And pushed the knapweed bunches where I stood
> When the mouse hurried from the crawling brood
> The young ones squeaked and when I went away
> She found her nest again among the hay
> The water oer the pebbles scarce could run
> And broad old cesspools glittered in the sun
>
> (PMS, A61-6)

Clare's feeling is closer to Burns's in his poem 'To a Mouse', yet
the poem is ostensibly descriptive. He leaves no room for complaint, but the final couplet unobtrusively echoes the plight of
the frightened mouse, not to mention the poet himself, staring
over the 'broad old cesspools'.

Clare was an ornithologist and botanist as painstaking as
Gilbert White. If he had had White's leisure and security there is
no doubt that there would have been a *Natural History of
Helpstone*, a project he had embarked upon. His Journal (1824-5)
is full of fascinating jottings:

Fri. 22 April 1825. Went to Milton—Saw the red-headed brown linnet
smaller than the brown do * —saw a Pettichap or hoverbird—& a large
flock of Fieldfares—brought home a white Primrose heard a many
Nightingales—in the evening I heard a bird make a long continued
noise for a minute together like a childs skriecker * or a cricket but
much louder—Henderson promises to give me some information res-
pecting the birds about Milton (*Prose* 144)

There is a moving entry for 9 March 1825, in which an empirical
observation suddenly takes over from a brooding on his own fate:

I had a very odd dream last night & I take it as an ill omen for I dont ex-
pect that the book will meet a better fate I thought I had one of the
proofs of the new poems from London & after looking at it awhile it
shrank thro my hands like sand & crumbled into dust the birds were
singing in Oxey Wood at 6 o clock this evening as loud & various as at
May (Ibid. 138)

Again he observes, 'How curious is the nest no other bird uses
such loose materials or weaves their dwellings in such spots—
dead oaken leaves are placed without and velvet moss within and
little scraps of grass'. The last quotation is not from a natural
history notebook or from the Journal; it is taken from a poem
called 'The Nightingales Nest', one of Clare's most beautiful
poems. To restore the quotation to its proper verse-form and to
continue to the end of the poem (the final 18 lines of a poem of
103 lines) is to witness natural history being transmuted into
poetry:

How curious is the nest no other bird
Uses such loose materials or weaves
Their dwellings in such spots—dead oaken leaves
Are placed without and velvet moss within
And little scraps of grass—and scant and spare
Of what seems scarce materials down and hair
For from mans haunts she seemeth nought to win
Yet nature is the builder and contrives
Homes for her childerns comfort even here
Where solitudes deciples spend their lives
Unseen save when a wanderer passes near
That loves such pleasant places—deep adown
The nest is made an hermits mossy cell
Snug lie her curious eggs in number five
Of deadened green or rather olive brown
And the old prickly thorn bush guards them well
And here we'll leave them still unknown to wrong
As the old woodlands legacy of song

(*SPP* 75)

It is clear that Clare's approach to nature is an endorsement of Blake's belief that 'He who would do good to another must do it in Minute Particulars'.[11] Blake and Clare both share a resistance to the neo-classical tendency to generalize, but they have little else in common. It is nearly always misleading to write of Clare's work in terms suitable for the other Romantics, as Harold Bloom does when he attempts to fit Clare into a Blakean framework (*The Visionary Company* (Ithaca 1971), pp. 444-56). It is difficult to understand what Bloom means when he writes, 'Clare does not imitate Wordsworth and Coleridge. He either borrows directly, or else works on exactly parallel lines ...' (p. 428). Bloom even calls 'To the Snipe' a 'Wordsworthian' poem, which is a vague misnomer (p. 430). He is equally misleading when he writes, 'Clare is the most genuine of poets, and yet it does not lessen him to say that much of his poetry is a postscript to Wordsworth's.' (p. 445.) It is my belief that to approach Clare on these terms is to lessen him considerably; in fact, it makes the early work appear incorrigibly minor and irrelevant. Clare must be seen on his own terms, or not at all.

Clare was both fascinated and repelled by Wordsworth. It is

remarkable that he was not more influenced by this contemporary giant. He writes in his Journal for Friday 29 October 1824:

When I first began to read poetry I dislikd Wordworth because I heard
he was dislikd & I was astonishd when I lookd into him to find my
mistaken pleasure in being delighted & finding him so natural & beautiful in his 'White Doe of Rylstone' there is some of the sweetest poetry I
ever met with tho full of his mysteries (*Prose* 118)

And in a manuscript fragment, Clare writes:

Wordsworth & Colridge, they are two favourites with me ... how like
you his Sonnet on 'Westminster Bridge' I think it (& woud say it to the
teeth of the critic in spite of his rule & compass) that it owns no equal in
the English language ... wordsworth defies all art & in all the lunatic
Enthuseism of nature he negligently sets down his thoughts from the
tongue of his inspirer but after all dont think I favour his affected
fooleries in some of his longer pieces theres some past all bearing[12]

These extracts show that Clare had read Wordsworth carefully,
and in his idiosyncratic way, had pin-pointed his strengths and
weaknesses. Clare, like many other critics, sees two voices in
Wordsworth: one speaking 'in all the lunatic Enthuseism of
nature' and 'full of his mysteries'; the other ponderous with
'affected fooleries'.

The key to Clare's uniqueness lies in his treatment of the
minute particulars of natural history. Wordsworth and Clare
both wrote sonnets about stepping-stones. A comparison of the
two poems will show that Clare does not imitate or borrow
directly from Wordsworth; his vision and his artistic aims are
quite different.[13] First, Wordsworth's sonnet:

> The struggling Rill insensibly is grown
> Into a Brook of loud and stately march,
> Crossed ever and anon by plank or arch;
> And, for like use, lo! what might seem a zone
> Chosen for ornament—stone matched with stone
> In studied symmetry, with interspace
> For the clear waters to pursue their race
> Without restraint. How swiftly have they flown,
> Succeeding—still succeeding! Here the Child

Puts, when the high-swoln Flood runs fierce and wild,
His budding courage to the proof; and here
Declining Manhood learns to note the sly
And sure encroachments of infirmity,
Thinking how fast time runs, life's end how near!
                    (*Wordsworth: PW* III. 249-50)

This is a poem of sentiment in the manner of Bowles or Rogers. No renovating 'spots of time' here, but only an awkward variation of the *sic transit* theme. Wordsworth seems more aware of the demands of the sonnet form than on the demands of the stepping-stones to be expressed; his eye is not on the object. The phrases 'ever and anon' and 'for like use, lo!' are vague, and help to pad out the ten-syllable line; the solemn words 'march', 'zone', and 'interspace' look uncomfortably as if they were there to supply the rhymes. One suspects that Clare would have described this sonnet as one containing 'affected fooleries'. Wordsworth's rejection of the Picturesque is not always for the best. Stepping-stones, unlike mountains, are not guaranteed to sustain the grandiloquence of his style. But, as Clare admits in his sonnet, stepping-stones are an admirably picturesque sight. Not only is his poem a series of pictures, it is also a series of pictures in motion. The picturesque eye is momentarily rivetted by detail, but the scene is not immobilized; the detail is involved in an animated environment:

The stepping stones that stride the meadow streams
Look picturesque amid springs golden gleams
Where steps the traveller with a wary pace
And boy with laughing leisure in his face
Sits on the midmost stone in very whim
To catch the struttles* that beneath him swim
While those accross the hollow lakes are bare
And winter floods no more rave dangers there
But mid the scum left where it roared and fell
The schoolboy hunts to find the pooty* shell
Yet there the boisterous geese with golden broods
Hiss fierce and daring in their summer moods
The boys pull off their hats while passing bye
In vain to fright—themselves being forced to flie[14]
                    (*SPP* 159)

The accuracy here is that of a painter. Even the repetition of 'golden' is not merely rhetorical, but an echo of local colour to balance the canvas. Yet the poem is by no means overburdened with detail or immobilized by painstaking accuracy like a Pre-Raphaelite painting, nor is it confined to an enumeration of picturesque scenes. Geese, in the way that they congregate and group, whether on land or on water, are a picturesque sight, but the line 'Hiss fierce and daring in their summer moods' goes far beyond the visual. The use of the word 'picturesque' in line 2 is not just an imprecise adjective, the equivalent of the debased modern usage of 'picturesque' as 'pretty'; it is used in precisely the way Uvedale Price had defined it:

In the same degree also, the simple construction of a foot-bridge ... formed by flat stones laid on more massy blocks, agrees with the character of a brook: indeed it generally happens that on a small scale, the rude efforts of inexperienced man have something more attractive, and what is very justly called picturesque, than that which is done by the more regular process of art ... (*Essays* II. 363)

It is important to remember that the picturesque theories of Price and Richard Payne Knight began as a reaction against the landscape gardening of Capability Brown, whose mid-eighteenth-century improvements imposed, in their own way, just as much design on nature as the French and Dutch gardens had done. Brown was translating Burke's concept of the Beautiful (smooth gradations) and Hogarth's Line of Beauty (the serpentine) into physical terms; his designs stress intellectual and structural qualities rather than effects of texture, variety, and intricacy. Price and Knight saw Brown's high-handed moulding of nature as tyranny and arrogance. Clare came to the same conclusion; his instinctive distrust of design, and his random readings, had enabled him to pick up enough knowledge to make judgements:

> The many whims that wealth requires
> Sloped mounts to peep at distant spires
> And bone paved grots where fancy tries
> How winter glooms suit summer skies
> And hermitage with wooden latch
> And ivy creeping oer the thatch

That might in her forgotten day
My Lady's pleasing taste display
Pretended love for simple things
A veil such as religion flings
Oft o'er disciples vague and hollow
To chuse a road they never follow
(PMS, B7-42b)

It is obvious from that extract that Clare had some knowledge of the landscapists. Indeed, he sometimes convinces one that he must have read Gilpin or Price at first hand. Whether he had or not, many of his poems read like catalogues of pictures by the Norwich school:

Old stonepits with veined ivy overhung
Wild crooked brooks oer which is rudely flung
A rail and plank that bends beneath the tread
Old narrow lanes where trees meet over head
Path stiles on which a steeple we espy
Peeping and stretching in the distant sky
And heaths oerspread with furze blooms sunny shine
Where wonder pauses to exclaim 'divine'
Old ponds dim shadowed with a broken tree
These are the picturesque of taste to me
While painting winds to make compleat the scene
In rich confusion mingles every green
Waving the sketchy pencil in their hands
Shading the living scenes to fairy lands
(*SPP* 160)

Inasmuch as Clare, with his partial knowledge of their theories, shares the reaction of Price and Knight against Brown, he is in the direct line of the picturesque movement. Cowper and Grahame, as might be expected, make similar protests against Brown's sweeping changes (*Cowper: PW*, p. 174; *The Birds of Scotland*, p. 46). Clare shares the picturesque theorists' delight in crumbling fabrics and in ruins, because ruins testify to nature's omnipotence and man's fragility, and he rejoices in this fact. It is his way of compensating for the manifold hardships he had suffered, many of them at the hands of his fellow-men. Uvedale Price had written in 1794:

I have shewn in an early part of my first Essay, how time and decay convert a beautiful building into a picturesque one, and by what process the change is operated. ... In proportion as the injury increases, in proportion as the embellishments that belong to architecture, the polish of its columns, the highly finished execution of its capitals and mouldings, its urns and statues are changed for what may be called the embellishments of ruins, for incrustations and weather stains, and for the various plants that spring from, or climb over the walls—the character of the picturesque prevails over that of the beautiful ... (*Essays* II. 258-9)

These sentiments are echoed all through Clare's work:

> There is a wild & beautiful neglect
> About the fields that so delights & cheers
> Where nature her own feelings to effect
> Is left at her own silent work for years
> (*MC* 452)

And Clare could be versifying Price when he writes about ivy:

> But bloom of ruins thou art sweet to me
>   When far from dangers way thy gloomy pride
> Wreaths picturesque around some ancient tree
>   That bows his branches by some fountain side
> There sweet it is from summer suns to be
> With thy green darkness overshadowing me.
> (NMS, 32-29)

Wordsworth also echoes Price's sentiments. We have seen how Wordsworth rejects the tyranny of the eye in *The Prelude*, but a poem such as 'Hart-Leap Well' is in the picturesque tradition, with its feeling for the 'bloom of ruins':

> 'The pleasure-house is dust:—behind, before,
>   This is no common waste, no common gloom;
> But Nature, in due course of time, once more
>   Shall here put on her beauty and her bloom.
>
> 'She leaves these objects to a slow decay,
>   That what we are, and have been, may be known;
> But at the coming of the milder day
>   These monuments shall all be overgrown.
> (*Wordsworth: PW* II. 254)

Both Wordsworth and Clare in their maturity translate picturesque sentiments into their own distinctive idiom, and they can
never be mistaken for one another. The Wordsworthian magic of
the line, 'But at the coming of the milder day' is out of Clare's
reach: in Clare's words, it is 'full of his mysteries'. On the other
hand Clare's detailed, fluid descriptions are not Wordsworth's
forte. When Robert Graves writes that Wordsworth, in comparison to Clare, 'had a very cursory knowledge of wild life; he
did not get up early enough in the morning',[15] he is not being
merely flippant—the distinction is worth bearing in mind in any
comparison between the two poets' visions.

In Wordsworth's poem 'The Sparrow's Nest', for example,
both the title and the first two lines could come straight from
Clare (with the exception of the admonishment 'Behold') but the
tone quickly changes:

> Behold, within the leafy shade,
> Those bright blue eggs together laid!
> On me the chance-discovered sight
> Gleamed like a vision of delight.
> I started—seeming to espy
> The home and sheltered bed,
> The Sparrow's dwelling, which, hard by
> My Father's house, in wet or dry
> My sister Emmeline and I
> Together visited.
>   (*Wordsworth: PW* I. 227)

This verse is a miniature demonstration of how Wordsworth's
mind works. The immediate details of the nest are forgotten as
he turns towards the effect the sight has on his mind; he characteristically converts it into 'a vision of delight', which is not the
same thing as saying, 'I am delighted by this particular nest in
front of me.' Wordsworth then turns to the past and to memories
of his father and of Dorothy's ministrations (the second verse
contains the much-quoted line 'She gave me eyes, she gave me
ears'), but on the evidence of this poem, sharpness of eye and ear
remain with Dorothy. William, in his male way, is forever ab-
stracting and pointing a moral. Dorothy's eye, as evidenced by
her *Journals*, is closer to Clare's.

Although Clare's vision and style are distinct from Wordsworth's, the two poets share indignation over man's misuse of
and tasteless interference with nature. Wordsworth's poem
'Effusion in the Pleasure-Ground on the Banks of the Bran, near
Dunkeld' (written in octosyllabic couplets like Clare's manuscript poem), castigates the 'baubles of theatric taste' on this particular estate, where a garden-house fitted up with mirrors gave
fantastic reflections of a waterfall:

> What pains to dazzle and confound!
> What strife of colour, shape and sound
> In this quaint medley, that might seem
> Devised out of a sick man's dream!
> . . . .
> Vain pleasures of luxurious life,
> For ever with yourselves at strife;
> Through town and country both deranged
> By affectations interchanged,
> And all the perishable gauds
> That heaven-deserted man applauds ...
>   (*Wordsworth: PW* III. 103, 105)

In the manuscript poem 'Walcott Hill and Surrounding Scenery', from which I have already quoted, Clare describes how
nature takes over from man's 'mad meddling toils', and the landscape reverts to wilderness.[16] Clare sees this process as natural
justice, and the words he uses are 'picturesque' and 'taste'. In the
following extracts, Clare delights in nature's tasteful and picturesque reappropriation of what was once a contrived picturesque
scene (* * denotes indecipherable word):

> Such * * heath thy wildered sight
> Gives taste a picturesque delight
> Though industrys mad meddling toils
> Thy wild seclusions yearly spoils
> Yet there are nooks still left behind
> As wild as taste could wish to find
> That toil has tried and tried in vain
> And left neglect its own again
> Which nature * * at freedom's will
> More sweet more wild and varied still

Staining in dies of greens and browns
Each little mountain's ups and downs
And climbing here and there to drop
Her wild thyme on each sandy top
And where each hollow streaks between
Spreading grass beds of deepening green
And snubbly bushes here and there
Where hares and rabbits love to lare
And sheep in summer eager run
To their short shadows from the sun
And where she finds a rock's edge bare
A huge stone careless lingering there
She sprinkles oer them passing by
Her witching tints of varied die
Mosses rust brown and green and gray
And when chance leaves them long from use
One's loath to see them mov'd away
Touched with her pencil's living hues
Such are wild heath thy hills and holes
                    (PMS, B7-40b, 41b)

This poem echoes Shaftesbury's delight in the 'genuine order' of informal nature, 'where neither Art, nor the Conceit or Caprice of Man has spoil'd their genuine order, by breaking in upon that primitive State'. Clare even echoes Shaftesbury's language, 'Quarries or camps of ancient times/ Where art broke on creation's sleep' (PMS, B7-38c). As the poem continues, Clare could be guiding us round Stowe:

And gravel walks by patches known
With weeds and grass nigh overgrown
These all in ruin's slumbers lie
And live the dream of what is by

....

A monument its column shows
Hill'd high in picturesque repose
Upon a gently-sloping mound
In memory of a favoured hound
Which time as heedless leaves alone
Nor hides the tale nor scars the stone
Nor cares as yet to climb the hill
But leaves his name to memory still

And here entombed poor 'Pompey' lies
And lessons still to pride supplies
Shows vanity its tinsel lot
That dogs can live when man's forgot
(PMS, B7-42b, 43b)

In his discussion of the eighteenth-century poetic garden, Ronald Paulson describes the Temple of British Worthies at Stowe (Pope's bust is at one end):

The Worthies are put into final, human perspective by one other reference. If you go around the back of the semicircular structure ... you find in the central arch (directly behind Mercury), a mock-heroic memorial to Signor Fido, one of Lord Cobham's dogs.[17]

The discreet neo-classical irony undercuts the formal tribute; Clare delights in bringing the irony to the foreground, and rejects man's vanity as 'tinsel'.[18]

Clare can learn from Wordsworth without being overpowered by him. He does not attempt to imitate him or borrow from him. He knows that Wordsworth is linguistically and philosophically in control of vast mountain panoramas, whereas he must look closer into the hedgerows and grasses to express the micro-panoramas made up of dynamic details. This narrowing of focus extends the picturesque vision into new territory.

'Adams Open Gardens'

We have seen how Clare comes to terms with picturesque vocabulary. We must now examine the problem of descriptive poetry.
This chapter illustrates some of the ways in which Clare solves
the problem of describing nature in his own way without falling
into dullness, sentimentality, or repetition. To create his own
system, like Blake, was beyond him, as was Henry Vaughan's
transcendental fervour, and he had no particular relish or gift for
metaphorical or symbolical profundities. His gift was for description and his subject-matter was the landscape. Descriptive
poetry, just as landscape painting, occupied a lowly rung of the
neo-classical ladder of perfection. Dr Johnson writes of Pope's
'Windsor-Forest':

The objection made by Dennis is the want of a plan, of a regular subordination of parts terminating in the principal and original design. There
is this want in most descriptive poems, because as the scenes, which
they must exhibit successively, are all subsisting at the same time, the
order in which they are shown must by necessity be arbitrary, and more
is not to be expected from the last part than from the first. The attention, therefore, which cannot be detained by suspense, must be excited
by diversity.[1]

The qualities specified as lacking are 'plan', 'subordination',
'design'—all words Clare had learned by bitter experience to
distrust. Keats wrote that he hated poetry that had 'a palpable
design upon us'; Clare suspected that where there was a 'regular
subordination' there was also 'a palpable design upon us'. Johnson's criticism is an acute but limited analysis of descriptive
poetry; it ignores the power of association and it assumes that the
whole can never be more than the sum of the parts.

A similar objection has been made often enough about Clare's
work. De Quincey wrote in 1840, 'The description is often true

even to a botanical eye; and in that, perhaps, lies the chief defect; not properly in the scientific accuracy, but that, in searching after this too earnestly, the feeling is sometimes too much neglected.' (*Heritage*, p. 246.) And Arthur Symons makes a complaint which can justifiably be levelled at much of Clare's less successful work, 'He begins anywhere and stops anywhere.'[2] Since lines of poetry have to be written successively—a poem like a piece of music moves through time, whereas the objects a descriptive poem sets out to delineate 'are all subsisting at the same time'—a problem of selection and emphasis immediately occurs. Only on one level, however, for as Proust has shown, our lives exist in time, but our inner history can be summoned immediately, in a state of what Blake called 'Organised Innocence', by a chance concurrence of the senses. The mature Clare uses synaesthesia and its power of association to enrich his visual images. He also breaks down the space-time perspective by his use of the forces of nature, such as storms, rain, snow, and mist. And he is nearly always in motion, walking through a landscape which is itself in the process of continual seasonal and atmospheric change. To walk through the open garden of nature involves a temporal and a visual sequence, but the most trivial event—a falling acorn, a travelling snail, the scent of beans in blossom, the barking of a dog in the middle-distance—might suddenly crystallize a thousand such walks into a cluster of images. The mature Clare enters a world, which to the casual eye may seem arbitrary, but to the imagination is freedom.

When perspective is eliminated, there is no subordination; all objects are of equal value, and therefore to talk of an arbitrary order is irrelevant. It is as if Clare is working on a medieval tapestry with a *mille fleurs* motif. The flowers are not numbered and placed; their glory is their profusion. They are *natura naturans*. We saw in Chapter One how the objects of Clare's world create their own dimensions instead of being contained in a spatial box, just as a peasant in an open-field system had a certain territorial freedom instead of being 'hedged about' by enclosure. The unenclosed landscape was Paradise for Clare, thus reversing the original meaning of Paradise, a word from the Persian meaning 'a walled enclosure'. The biblical Fall is also reversed—instead of being ejected from a Garden, Clare loses 'Adams open gardens' (*SPP* 184) as they are 'subordinated' to enclosure.

Johnson censures Pope for relying too much upon description. But Pope himself, in his 'Discourse on Pastoral Poetry', had written of Spenser's *The Shepheardes Calender* that

> ... the scrupulous division of his Pastorals into Months, has oblig'd him either to repeat the same description, in other words, for three months together; or when it was exhausted before, entirely to omit it: whence it comes to pass that some of his Eclogues (as the sixth, eighth, and tenth for example) have nothing but their Titles to distinguish them. The reason is evident, because the year has not that variety in it to furnish every month with a particular description, as it may every season.
>
> (*Pope: Poems* I. 32)

Whether or not that statement is true of Spenser's poem, it goes very wide of the mark if applied to Clare's *The Shepherd's Calendar*. As a countryman who had been obliged to labour in all weathers, Clare knew that the English year has an almost infinite variety of moods, and that every week has a different flavour, let alone every month. But a challenge to Pope's neo-classical view of natural description had been in existence long before Clare's day. Joseph Warton, in his *An Essay on the Writings and Genius of Pope* (1756), writes, 'MR. POPE it seems was of opinion, that descriptive poetry is a composition as absurd as a feast made up of sauces: and I know many other persons that think meanly of it.'[3] Of the Pastorals, he has this to say:

> It is something strange, that in the pastorals of a young poet there should not be found a single rural image that is new: but this I am afraid is the case in the PASTORALS before us. (Ibid. 2)

Warton goes on to hold Thomson up as a model as opposed to Pope:

> Thomson was blessed with a strong and copious fancy; he hath enriched poetry with a variety of new and original images, which he painted from nature itself, and from his own actual observations: his descriptions have therefore a distinctness and truth, which are utterly wanting to those, of poets who have only copied from each other, and have never looked abroad on the objects themselves. Thomson was accustomed to

wander away into the country for days and for weeks, attentive to 'each rural sight, each rural sound.' (Ibid. 42)

Warton uses the phrase 'he painted from nature itself'; the whole passage, indeed, with the substitution of a few words, might be a later critic's description of Constable. Warton continues, and specifies passages worthy of admiration:

Innumerable are the little circumstances in his descriptions, totally unobserved by all his predecessors.... How full, particular and picturesque is this assemblage of circumstances that attend a very keen frost in a night of winter!

> Loud rings the frozen earth, and hard reflects
> A double noise; while at his evening watch
> The village dog deters the nightly thief;
> The heifer lows; the distant water-fall
> Swells in the breeze; and with the hasty tread
> Of traveller, the hollow-sounding plain
> Shakes from afar.
>
> (Ibid. 43-5)

John Aikin also denigrates Pope, and has high praise for Thomson, whose plan is 'scarcely less extensive than nature itself...' (*Essay*, p.59). Aikin reiterates Joseph Warton's praise of one of Thomson's passages and also uses the word 'picturesque':

A striking instance of the extraordinary effect of a well-chosen epithet in adding life and force to a description, is shewn in the expression '*buzzing* shade.' A single word here conveys to the mind all the imagery of a passage in the same author which Mr. Warton justly commends as equally new and picturesque.

'Resounds the living surface of the ground' ... It is by means of such bold comprehensive touches as these, that Poetry is frequently enabled to produce more lively representations than Painting, even of sensible objects. (Ibid. 72-3)

The passages which Warton and Aikin single out for praise are nearly all auditory ones, and both critics use the word 'picturesque' in describing the power of these passages. Later critics do likewise. In a review of *The Shepherd's Calendar*, Josiah Conder writes, 'The opening lines of "May" would form a good subject for Wilkie, were it not that painting cannot be so picturesque as language, which can express, as Dugald Stewart remarks, picturesque sounds as well as sights, and picturesque sentiments also.' (*Heritage,* p. 204.) 'May' begins as follows:

> Come queen of months in company
> Wi all thy merry minstrelsy
> The restless cuckoo absent long
> And twittering swallows chimney song
> And hedge row crickets notes that run
> From every bank that fronts the sun
> And swathy* bees about the grass
> That stops wi every bloom they pass
> And every minute every hour
> Keep teazing weeds that wear a flower
> And toil and childhoods humming joys
> For there is music in the noise
> The village childern mad for sport
> In school times leisure ever short
> That crick* and catch the bouncing ball
> And run along the church yard wall
> ....
> And jilting* oer the weather cock
> Viewing wi jealous eyes the clock[4]

In a manuscript fragment of a novel, Clare expresses picturesque sounds, sights, and sentiments together:

There was a small stream went bending in roundabout mazes across the forest and now and then interrupted by a sallow* bush that longing to kiss the water had bent its mossy fringed and shaggy rooted branches not only over but into the flood and from these interruptions and the broken down pathways of stones laid by rustics for the benefit of nearer way the waters broke their silence into the beautiful murmuring music that one often hears on suddenly approaching these picturesque rivulets and suddenly lose when we leave their banks[5]

The Shepherd's Calendar, similarly, is a series of precise genre pictures, given unusual immediacy and vitality by a poet who experiences natural phenomena with five alert senses. Where the neo-classical poet tends to downgrade the other senses to give primacy to vision, eighteenth-century descriptive poets rely more and more on the evocations of synaesthesia. The senses enhance each other, as in Thomson's lines 'Along the soft-inclining fields of corn' (*Thomson: PW*, p. 144), and 'Now, while I taste the sweetness of the shade' (ibid., p. 77).

In 'A Nocturnal Reverie' (1713), a favourite poem of Wordsworth's, Lady Winchilsea subdues the usual features of the landscape, and evokes sounds and smells which construct a perspective of their own. The well-known line about the horse, 'Till torn up Forage in his Teeth we hear', is only one sound among others—the wind, Philomel, falling waters, curlews, the partridge—which space the landscape out in the absence of visual clarity. Lady Winchilsea imparts a powerful unity of effect, but daylight and the visual disturb the harmony, 'Till Morning breaks, and All's confus'd again'.

In Gray's Elegy (1751), as the 'glimmering Landscape' fades from sight, it is the sounds, such as the 'drowsy Tinklings', which set the scene and construct the landscape. This aural spacing is more immediate, more sensuous, and more nostalgic, and it is with a sense of joyful collaboration with the darkness, comparable to Lady Winchilsea's, that Gray writes, 'And leaves the World to Darkness, and to me.'

Clare does not employ Thomson's elaborate syntax or Gray's and Lady Winchilsea's league with darkness in order to experience nature directly. Like theirs, his acute visual sense is active, but it is always enhanced and enriched by the other senses:

> The south west wind how pleasant in the face
> It breathes while sauntering in a musing pace
> I roam these new ploughed fields and by the side
> Of this old wood where happy birds abide
> And the rich blackbird through his golden bill
> Litters wild music when the rest are still
> Now luscious comes the scent of blossomed beans
> That oer the path in rich disorder leans
> Mid which the bees in busy songs and toils
> Load home luxuriantly their yellow spoils
> The herd cows toss the molehills in their play
> And often stand the strangers steps at bay
> Mid clover blossoms red and tawney white
> Strong scented with the summers warm delight
>
> (*SPP* 159)

Clare's eye is sharp, but not exclusive; it plays an essential but co-operative role in his complete sensuous response to landscape. This co-operation is much in evidence in the following prose fragment (c. 1847):

The rustling of leaves under the feet in woods and under hedges
The crumping of cat ice and snow down wood-rides, narrow lanes and every street causeway
Rustling thro a wood or rather rushing, while the wind hallows in the oak tops like thunder;
The rustle of birds' wings startled from their nests or flying unseen into the bushes
The whizzing of larger birds overhead in a wood, such as crows, pud-docks*, buzzards, Ec,
The trample of robins & woodlarks on the brown leaves, and the patter of squirrels on the green moss;
The fall of an acorn on the ground, the pattering of nuts on the hazel branches as they fall from ripeness;
The flirt* of the ground-lark's wing from the stubbles—how sweet such pictures on dewy mornings when the dew flashes from its brown feathers!

(Prose 251)

It is characteristic that Clare should end by saying 'how sweet such pictures' instead of 'how sweet such sounds'. The fact that he does so sharpens the sounds by giving them a visual setting, and avoids a mere list.

It is difficult for the descriptive poet to get far without solving the problem of list passages. With Milton's sonorous flower piece in 'Lycidas' in mind, compare Thomson and Clare as botanists. First, Thomson:

> But why so far excursive? when at hand,
> Along these blushing borders bright with dew,
> And in yon mingled wilderness of flowers,
> Fair-handed Spring unbosoms every grace—
> Throws out the snow-drop and the crocus first,
> The daisy, primrose, violet darkly blue,
> And polyanthus of unnumbered dyes;
> The yellow wall-flower, stained with iron brown,
> And lavish stock, that scents the garden round:

From the soft wing of vernal breezes shed,
Anemones; auriculas, enriched
With shining meal o'er all their velvet leaves;
And full ranunculus of glowing red.
Then comes the tulip-race, where beauty plays
Her idle freaks: from family diffused
To family, as flies the father-dust,
The varied colours run; and, while they break
On the charmed eye, the exulting florist marks '
With secret pride the wonders of his hand.

(*Thomson: PW* 23)

That passage is never dull, because Thomson varies the pauses, rhythms, and emphases so subtly. Without this careful punctuation, such a passage might well fall into monotony. In the following passage from 'June', Clare does not rely upon punctuation, and many of his lines begin with 'And'. But he has evolved a rhythm of his own, with its slightly awkward countryman's gait; the strength of such awkwardness is evident in such a line as 'Their honey-comb-like blossoms hanging down'. And behind Clare's lines echoes the vast impersonal world of folk-songs and balladry, with its jilted maidens and blossoming, withering flowers. Clare's lines are less accomplished than Thomson's, but they have an urgency and a poignancy which arise from Clare's knowledge of his loss of Eden:

Old fashiond flowers which huswives love so well
And columbines stone blue or deep night brown
Their honey-comb-like blossoms hanging down
Each cottage gardens fond adopted child
Tho heaths still claim them where they yet grow wild
Mong their old wild companions summer blooms
Furze brake and mozzling* ling and golden broom
Snap dragons gaping like to sleeping clowns
And 'clipping pinks'* (which maidens sunday gowns
Full often wear catcht at by tozing* chaps)
Pink as the ribbons round their snowy caps
'Bess in her bravery' too of glowing dyes
As deep as sunsets crimson pillowd skyes
And majoram notts sweet briar and ribbon grass
And lavender the choice of every lass

And sprigs of lads love * all familiar names
Which every garden thro the village claims
These the maid gathers wi a coy delight
And tyes them up in readiness for night
                    (*SC* 67-8)

This passage has a story-teller's skill, where repetition can
become almost hypnotic, but is never allowed to become trite.
Clare wants to convey the sensuous plenitude of rustic life and is
not concerned with the 'regular subordination of parts terminating in the principal and original design'. Furthermore,
such passages are nourishing meals in themselves, not a 'feast
made up of sauces', concocted to distract attention from a central
lack of substance. If one compares *The Shepherd's Calendar* with
any of its predecessors (Mark Storey has shown how Clare was
consciously working in a well-established tradition),[6] one sees
that to call Clare a 'descriptive' poet or a 'nature' poet is misleading.

One of the few poets before Clare to divide the year into twelve
months, instead of four seasons, was James Grahame. *The Rural
Calendar* is full of delicate and closely-observed touches, but the
space allotted to each month is very small, and much of that
space is taken up with genre scenes. The following quotation is
Grahame's 'September' in full. This is a countryman's observation, but Grahame gives the impression that he finds it difficult
to differentiate each month sufficiently, and Pope might have
censured *The Rural Calendar* for repetition:

Gradual the woods their varied tints assume;
The hawthorn reddens, and the rowan-tree
Displays its ruby clusters, seeming sweet,
Yet harsh, disfiguring the fairest face.

At sultry hour of noon, the reaper band
Rest from their toil, and in the lusty stook
Their sickles hang. Around their simple fare,
Upon the stubble spread, blythesome they form
A circling groupe, while humbly waits behind
The wistful dog, and with expressive look,
And pawing foot, implores his little share.

The short repast, seasoned with simple mirth,
And not without the song, gives place to sleep.
With sheaf beneath his head, the rustic youth
Enjoys sweet slumbers, while the maid he loves
Steals to his side, and screens him from the sun.

But not by day alone the reapers toil:
Oft in the moon's pale ray the sickle gleams,
And heaps the dewy sheaf;—thy changeful sky,
Poor Scotland, warns to seize the hour serene.

The gleaners, wandering with the morning ray,
Spread o'er the new-reaped field. Tottering old age,
And lisping infancy, are there, and she
Who better days has seen.—
                No shelter now
The covey finds; but, hark! the murderous tube.
Exultingly the deep-mouthed spaniel bears
The fluttering victim to his master's foot:

Perhaps another, wounded, flying far,
Eludes the eager following eye, and drops
Among the lonely furze, to pine and die.
                        (*Grahame: Poems* I, 113-14)

Grahame's 'September' is descriptive in the sense that it is a verbal sketch which relies almost exclusively on the poet's eye—an eye, moreover, accustomed to seeing things from the point of view of traditional perspective, no matter how rustic the subject-matter. The wounded bird eludes the 'eager following eye' because the poet is necessarily limited to one point of view, to a linear perspective analogous to the rational progression of his carefully punctuated language. Clare's vision, on the other hand, is akin to that of the Zen master who tweaked his student's nose for saying 'the birds have flown away'. They fly away only if we see them from one static viewpoint, whereas Clare's eye is continually moving with them, now a hundred feet in the air, now in the bushes.

When Grahame wants to describe sounds he is rather at a loss: 'hark! the murderous tube'. In Clare's *The Shepherd's Calendar*,

all the senses are brought into play, the visual images 'earthed'
and made more immediate by tactile and auditory ones. The
following is an extract from Clare's 'September':

> Loud are the mornings early sounds
> That farm and cottage yard surrounds
> The creaking noise of opening gate
> And clanking pumps where boys await
> With idle motion to supply
> The thirst of cattle crowding bye
> And low of cows and bark of dogs
> And cackling hens and wineing hogs
> Swell high—while at the noise awoke
> Old goody seeks her milking cloak
> And hastens out to milk the cow
> And fill the troughs to feed the sow
> Or seeking old hens laid astray
> Or from young chickens drives away
> The circling kite that round them flyes
> Waiting the chance to seize the prize
> Hogs trye thro gates the street to gain
> And steal into the fields of grain
> From nights dull prison comes the duck
> Waddling eager thro the muck
> Squeezing thro the orchard pales
> Where mornings bounty rarely fails
> Eager gobbling as they pass
> Dew worms thro the padded* grass
>   (SC 105-6)

In *The Problem of Style*, Middleton Murry writes that Clare
possesses 'the true poetic activity ... in its least complicated form'
(London, 1960, p. 90). He describes this as the appropriate
crystallization of emotion around an object, a perfect fusion of inner and outer, 'The sensuous perceptions have aroused an emotional apprehension ... objects being in an active relation to the
emotion, the emotion is crystallized about them' (ibid., p. 94).
This fusion is much less likely to occur when the description is
purely visual, as objects have a more passive relationship to the
onlooker. To 'describe' a landscape, then, as Clare does, is to see
it, feel it, hear it, smell it, and taste it. It is to acknowledge its

strange otherness while being at the same time a part of it. This 'description' is quite different from the tradition described in Chapter One in which the poet is able to 'command' a landscape by distancing it visually. Even Cowper resorts to this military metaphor, 'Now roves the eye;/ And, posted on this speculative height,/ Exults in its command' (*Cowper: PW*, p. 135).

This visual distancing is taken for granted by Dr Johnson when he writes of the 'regular subordination of parts terminating in the principal and original design'. To 'design' a landscape in this way is to make most of it passive while the poet's eye 'roves'. A descriptive poem of this sort easily deteriorates into a catalogue of static details, where 'the order in which they are shown must by necessity be arbitrary'. But Clare's best descriptions, by always maintaining an 'active relation' between subject and object, are never just visual inventories. By eschewing a single point of view, Clare can capture perfectly the individual perceptions of different labourers, where pain and pleasure are mingled inextricably:

> The hedger toils oft scaring rustling doves
> From out the hedgrows who in hunger browze
> The chockolate berrys on the ivy boughs
> And flocking field fares speckld like the thrush
> Picking the red awe * from the sweeing * bush
> That come and go on winters chilling wing
> And seem to share no sympathy wi spring
> The stooping ditcher in the water stands
> Letting the furrowd lakes from off the lands
> Or splashing cleans the pasture brooks of mud
> Where many a wild weed freshens into bud
> And sprouting from the bottom purply green
> The water cresses neath the wave is seen
> Which the old woman gladly drags to land
> Wi reaching long rake in her tottering hand
> The ploughman mawls * along the doughy sloughs
> And often stop their songs to clean their ploughs
> From teazing twitch that in the spongy soil
> Clings round the colter terryfying toil
> The sower striding oer his dirty way
> Sinks anckle deep in pudgy * sloughs and clay

And oer his heavy hopper stoutly leans
Strewing wi swinging arms the pattering beans
Which soon as aprils milder weather gleams
Will shoot up green between the furroed seams

                                        (*SC* 31)

There are obvious sounds in this passage ('rustling doves' and 'pattering beans'), but a dialect word such as 'pudgy' adds greatly to the vividness of impression—an example of how a dialect word can be more precise and true to experience. Writing on dialect, Melvyn Bragg quotes Hardy, 'Dialect words—those terrible marks of the beast to the truly genteel'. Bragg could be speaking on Clare's behalf when he writes, 'Also I have the fear that if you lose the name of the thing you lose the thing itself. The naming of parts took a long time and it is a great pity that we should neglect what was so hard to come by.' (*The Listener*, 19 January 1975, pp. 52, 53.) Elevated language tends to assimilate details into the naming of wholes. Clare's recalcitrance on the matter of dialect stems from a fear similar to Bragg's, but more urgent. Clare is dedicated to the naming of parts because local details, his minute particulars, make up the world as he can experience it and love it. Passages like the above abound in Clare's best work, where a dramatic sense of movement provides changes of tempo and focus which can only be called cinematic. In the line 'And flocking field fares speckld like the thrush', the eye takes in simultaneously the rapid swoop of the flock in flight and the close-up detail of the birds' markings. And in line 11 of 'The Moorehens Nest', Clare writes, 'I pick out pictures round the fields that lie'; he then goes nesting, and line 67 reads quite differently, 'And catch at little pictures passing bye' (*SPP*, pp. 79, 81). In line 11 the pictures are static; in line 67 the cinematograph of nature has set the thousands of individual static frames in motion.

In spite of, or perhaps because of, these qualities, *The Shepherd's Calendar* sold very poorly when it finally appeared in 1827. The emerging fondness for still life (*nature morte*) was inimical to this kind of vision. It was a long time before poetry recovered it. In 1905, W. B. Yeats complained, 'I think the whole of our literature as well as our drama has grown effeminate through the over development of the picture-making faculty. The great thing in literature, above all in drama, is rhythm and movement.'[7] Just as the medieval world feared 'the lust of the eye' and Wordsworth reacted against the eye's tyranny, modern poets have passed through the vanishing point and created spaces of their own. In this respect Clare, like Hopkins, should be seen as an important precursor. Clare's originality arises from the way in which he blends what I have called an 'insect view' with a strong literary intelligence. One of the hallmarks of this intelligence is a characteristically rural virtue: patience. To an unsympathetic outsider, patience can look like stupidity, but Clare's enormous patience, coupled with a sharp eye and an alert intelligence, enabled him to participate in nature in a way which very few artists or poets have ever done. This participation transcended what Yeats calls a 'picture-making faculty' to achieve an understanding of what a modern naturalist, commenting on an illustration of the teeming life of a typical English hedgerow, calls a 'dynamic picture':

In reality you will not see at any one moment as much as we show within the few square yards or metres in our drawing. But you might if you watched long enough. There lies the first lesson: a plant or animal is playing a part in a chain of events, which you could not guess from a single sighting of it. Everything you see is part of a bigger, dynamic picture. (*The Sunday Times Book of the Countryside* (London, 1980), p. 90.)

In the following extract from 'October', one can see a series of sharply observed events merge until one gets a sense of 'a bigger, dynamic picture'. It is as if the individual frames of each single-sighted perception 'fade' into each other so that nature is seen as a continuum:

> The cotter* journying wi his noisey swine
> Along the wood side where the brambles twine
> Shaking from dinted cups the acorns brown
> And from the hedges red awes* dashing down
> And nutters rustling in the yellow woods
> Scaring from their snug lairs the pheasant broods
> And squirrels secret toils oer winter dreams

Picking the brown nuts from the yellow beams
And hunters from the thickets avenue
In scarlet jackets startling on the view
Skiming a moment oer the russet plain
Then hiding in the colord woods again
The ploping guns sharp momentary shock
Which eccho bustles from her cave to mock
The sticking groups in many a ragged set
Brushing the woods their harmless loads to get
And gipseys camps in some snug shelterd nook
Where old lane hedges like the pasture brook
Run crooking as they will by wood and dell
In such lone spots these wild wood roamers dwell
On commons where no farmers claims appear
Nor tyrant justice rides to interfere
Such the abodes neath hedge or spreading oak
And but discovered by its curling smoak
Puffing and peeping up as wills the breeze
Between the branches of the colord trees
Such are the pictures that october yields
To please the poet as he walks the fields

(*SC* 112-13)

One of the strengths of this passage is the fact that Clare is in
motion 'as he walks the fields' in the open garden of nature. In
eighteenth-century gardens (The Leasowes, Rousham, Stowe,
Stourhead), one of the main features was the perimeter walk.
In following this winding circular path, one could participate
in the whole landscape as one's shifting viewpoint provided
an almost endless series of prospects and retrospects. The invention of the ha-ha or sunken fence allowed the garden to continue visually beyond the perimeter. Horace Walpole credits
Charles Bridgeman with the invention of the ha-ha but William
Kent with its first effective use. Kent 'saw that all nature was a
garden'.[8] John Dixon Hunt has shown how techniques learned
inside the landscape garden were applied later in the century
to the whole landscape; the sense of discovery which visitors
always experienced in these gardens (the name ha-ha came from
the exclamations of surprise on discovering the hidden fence)
was extended to nature as a whole, 'Poetry that is organized after

the pattern of landscape gardens is characterized by its concern with movement through scenery, the "kinema" not the prospect picture, the mysterious process of discovering what comes next to eye and mind rather than an apprehension of landscape at a distance...' (*The Figure in the Landscape*, p. 228). Clare is continually exclaiming 'ha-ha' and leaping visual fences in order not just to see, but to experience all nature as a garden. His experience of landscape becomes a continuous series of elements melting into each other.

It is Clare's use of the forces of nature as active collaborators with the mind in enlivening and transfiguring the landscape which most distinguishes him from eighteenth-century theory. It was of course Thomson, as Johnson, Warton, and Aikin were aware, who first exploited to the full the accidental and ephemeral aspects of nature to make his poetic effects. As we have seen, both knowledge of natural history and interest in the Picturesque had increased enormously since the publication of *The Seasons* (1726-30). The development of the Picturesque is a progression from the mind as passive spectator of a preconceived space to the mind as active participant in a mysterious intercourse with natural forces. Uvedale Price writes of the Picturesque with the Wordsworthian phrase 'active agency'.[9] Wordsworth himself, in a cancelled manuscript fragment of *The Prelude*, which could none the less stand as epigraph to that poem, writes, 'Nature works/Herself upon the outward face of things/As if with an imaginative power' (*The Prelude*, ed. Ernest de Selincourt (Oxford, 1959), p. 623). Although Clare rejects the narrower aspects of the Picturesque, he harnesses this 'active agency' to his own advantage. He creates landscapes which are dynamically selective, not just randomly descriptive.

An interest in the processes of nature rather than its forms, in essences rather than appearances, in weather rather than topography, is one of the most marked facets of English Romanticism (description and topography are not, of course, absent from the work of the Romantics—Scott's feeling for locality, Keats's sensuous apprehension of detail, and much of Wordsworth's work, such as 'Poems on the Naming of Places', provide evidence to the contrary). This search for the essence is nothing vague. It is accompanied by an acute feeling for minutiae, a feeling nurtured by the growing sophistication of natural history and by the rigorous training of the eye promoted by the picturesque theorists. The ideal is to be alive to a vast number of stimuli, but then to be able to fuse them into significant wholes. This control of 'inner weather' is for Wordsworth and Coleridge the faculty of Imagination; it is seen pre-eminently at work in such poems as 'Frost at Midnight', 'Tintern Abbey', and *The Prelude*, in which the 'auxiliar light' of the mind takes many of its metaphors from atmospheric conditions. The great Romantic painters have similar beliefs about the imaginative process. Constable defends one of his house portraits against numerous complaints by describing it as 'a picture of a summer morning, *including a house*', thus epitomizing the transition from topography to vision.[10] And Turner, who loved Thomson's work and who sometimes quotes from *The Seasons* on the back of a canvas, 'came to realize that the *forms of movement* were what he wanted to define, and that nature consisted, not of separate objects in mechanical relations to one another, but of fields of force'.[11]

An imaginative co-operation with weather is evidenced by the following entry in Dorothy Wordsworth's 'Alfoxden Journal':

1st March. We rose early. A thick fog obscured the distant prospect entirely, but the shapes of the nearer trees and the dome of the wood dimly seen and dilated. It cleared away between ten and eleven. The shapes of the mist, slowly moving along, exquisitely beautiful; passing over the sheep they almost seemed to have more of life than those quiet creatures. The unseen birds singing in the mist.[12]

The prospect is obscured and objects are transformed and 'dilated'. The mist takes on a life of its own and 'unseen birds' construct aural spaces with their song. Compare Clare's sonnet, 'Mist in the Meadows':

> The evening oer the meadow seems to stoop
> More distant lessens the diminished spire
> Mist in the hollows reaks * and curdles up *
> Like fallen clouds that spread—and things retire
> Less seen and less—the shepherd passes near
> And little distant most grotesquely shades

As walking without legs—lost to his knees
As through the rawky* creeping smoke he wades
Now half way up the arches dissappear
And small the bits of sky that glimmer through
Then trees loose all but tops—while fields remain
As wont—the indistinctness passes bye
The shepherd all his length is seen again
And further on the village meets the eye
(*SPP* 154-5)

This poem begins with a personification in the manner of the eighteenth century but immediately passes into a meditation on the processes of nature and how the eye is inadequate when faced with them. The agency of mist makes nature an artist with grotesque and surrealistic gifts, arbitrary and independent. There is nothing sinister about it, as there are sinister forces in Clare's Asylum verse, but it is indifferent to human perspectives. Clare rejoices in this, and makes us feel at the end that the shepherd who has found his feet again and the village emerging from the mist have been through a mysteriously purgative process. Clare's ostensibly descriptive verse nearly always adds up to more than the sum of its parts and gives the reader a feeling of wholeness which a mere visual response could not. This feeling can perhaps only be accounted for in extra-literary terms.

In a remarkable book, *The Experience of Landscape* (London, 1975), Jay Appleton proposes an approach to landscape, in all its manifestations, by the adoption of prospect-refuge symbolism and what he calls 'habitat theory'. He argues that our physical, visual, and imaginative experiences of landscape are much more closely integrated than is usually allowed, and that our aesthetic delight in a prospect, for example, is directly related to the hunter's experience of stalking his prey, where success (and often survival) depend upon a strategic use of the environment—in Konrad Lorenz's phrase, 'to see without being seen':

Habitat theory postulates that aesthetic pleasure in landscape derives from the observer experiencing an environment favourable to the satisfaction of his biological needs. Prospect-refuge theory postulates that, because the ability to see without being seen is an intermediate step in the satisfaction of many of those needs, the capacity of an environment to ensure the achievement of *this* becomes a more immediate source of aesthetic satisfaction. (Ibid. 73)

Whatever the validity of this theory (and Appleton, in spite of marshalling an impressive amount of supporting evidence, remains diffident),it must be admitted that it casts a great deal of light upon Clare's work. Appleton writes:

A field mouse can find concealment in quite short grass which would afford us no protection whatever, and as we watch him extracting the maximum advantage from his tiny world of prospects and refuges we momentarily live his life for him ... (Ibid. 190)

I suggested in Chapter Two that Gilbert White was in a position to be detached about the fate of a field-mouse, whereas Clare reacts in a strangely intense, personal way. What else is Clare doing but momentarily living the life of this hunted animal as it strategically employs its 'tiny world of prospects and refuges' for survival? He does the same thing in nearly all his bird poems, especially the ones dealing with those strong prospect-refuge symbols, birds' nests:

It nestled like a thought forgot by toil
& seemed so picturesque a place for rest
I een dropt down to be a minutes guest
& as I bent me for a flower to stoop
A little bird cheeped loud & fluttered up
The grasses tottered with their husky seeds
That ramped * beside the plough with ranker weeds
I looked—& there a snug nest deep & dry
Of roots & twitches * entertained my eye
& six eggs sprinkled oer with spots of grey
Lay snug as comforts wishes ever lay
The yellow wagtail fixed its dwelling there
Sheltered from rainfalls by the shelving share
That leaned above it like a sheltering roof
From rain & wind & tempest comfort proof
Such safety-places little birds will find
Far from the cares & help of human kind
For nature is their kind protector still
To chuse their dwellings furthest off from ill

> So thought I—sitting on that broken plough
> While evenings sunshine gleamed upon my brow
> So soft so sweet—& I so happy then
> Felt life still eden from the haunts of men
>
> (*MC* 212)

Much of Clare's extensive knowledge of natural history is acquired by a process of obsessive identification:

I have often wondered how birds nests escape injuries which are built upon the ground I have found larks nests in an old cart rut grassed over and pettichaps* close on the edge of a horse track in a narrow lane where two carts could not pass and two oxen would even have difficulty in doing so but yet I never found a nest destroyed (*SPP* 108)

This process is described by Appleton, 'From the vicarious participation in landscape through identification with the wild animal it is but a short step to the pathetic fallacy ... the whole environment is translated from a passive to an active role.' (*The Experience of Landscape*, p. 191.) This helps to account for the dynamic quality of Clare's nesting poems:

> Here did I roam while veering over head
> The pewet whirred in many whewing rings
> And 'chewsit' screamed and clapped her flopping wings
> To hunt her nest my rambling steps was led
> Oer the broad baulk* beset with little hills
> By moles long formed and pismires* tennanted
> As likely spots but still I searched in vain
> When all at once the noisey birds were still
> And on the lands a furrowed ridge between
> Chance found four eggs of dingy olive green
> Deep blotched with plashy spots of jockolate stain
> Their small ends inward turned as ever found
> As though some curious hand had laid them round
>   . . . .
> Hid from all sight but the allseeing sun
> Till never ceasing danger seemeth bye
>
> (*SPP* 82-3)

Clare's constant awareness of danger marks him off from much Romantic nature poetry. Wordsworth is as aware of danger as of beauty, but Keats's Nightingale and Shelley's Skylark are symbols of an ideal state of mind, not creatures struggling with their environment to survive. This aspect of Clare has often been overlooked by critics, and Middleton Murry even writes that Clare ignores the violence of nature.[13] But L. J. Swingle's comments are much closer to our present theme, 'Clare's creatures scurry about through his descriptive poetry, sometimes destroying and being destroyed, but more often merely treading the brink of destruction and managing, for the moment, to escape.... Birds may sing like angels, but they hunger and they fear like men.'[14] According to habitat theory birds (and poets) sing like angels *because* they hunger and fear. And Clare's fascination with molehills obviously goes beyond the fact that they mimic Parnassus; they are also potent refuge-symbols, as they remind him that the mole is safely hidden in the ground. Furthermore, habitat theory may help to explain Clare's dislike of the Capability Brown style, in that it is difficult to hide in it (Christopher Hussey explains that the Le Nôtre style of geometrical alleys, as at Versailles, was designed to facilitate hunting).[15] Given the fact that Clare identified strongly with hunted animals and birds, it is not surprising that he finds such landscapes repellent. The picturesque writers sought a quite different landscape, and Clare, as we have seen, is a picturesque writer inasmuch as he shares their passion for roughness, irregularity, and the 'bloom of ruins'. Again, habitat theory is relevant, 'All kinds of irregularity tend to facilitate the act of hiding. Rough, broken ground, through which one may move easily from one boulder to another, provides a more effective cover than a much larger feature, such as a cliff, which denies access.' (*The Experience of Landscape*, p. 105.)

In Thomas Bewick's *A Memoir*, the word 'picturesque' is used to describe a scene on the old heath which Bewick knew as a boy. Clare's feeling for 'wild seclusions' finds an ideal counterpart in Bewick's prose:

He kept a few Sheep, on the Fell & in pretence of looking after these, his secret & main business was looking after his Bees;—of these he had a great number of Hives, all placed in very hidden & very curious situations—the narrow entrances to all of these were at some distance from the Hives—some of these Apiaries were placed under the bounder hedge of the common, surrounded & hidden from sight by a great extent of old whin bushes, and besides, the Hives were sheltered under the branches of the old thorns and almost quite covered or over hung by Brambles, Woodbine & hip Bryars, which, when in blossom, looked beautifully picturesque, while at the same time they served to keep the eye from viewing the treasures thus concealed beneath.[16]

When Clare cannot take cover behind vegetation, in a wood or in long grass, he attempts to find refuge in other ways—by constant movement, by resorting to that micro-panoramic 'world of prospects and refuges', or by calling upon the forces of nature to impede visibility.

We are now in a better position to examine Clare's use of natural forces, which offer, apart from variety and movement, the chance of concealment. Clare's intense visual delight in mist is enhanced by the fact that the mist allows him a good strategic position for observation without making him vulnerable (as a thick fog would). His involvement with the forces of nature has an atavistic alertness which activates the keen visual details, as in another sonnet about mist:

> The shepherds almost wonder where they dwell
> And the old dog for his night journey stares
> The path leads somewhere but they cannot tell
> And neighbour meets with neighbour unawares
> The maiden passes close beside her cow
> And wonders on and think[s] her far away
> The ploughman goes unseen behind his plough
> And seems to loose his horses half the day
> The lazy mist creeps on in journey slow
> The maidens shout and wonder where they go
> So dull and dark are the november days
> The lazy mist high up the evening curled
> And now the morn quite hides in smokey haze
> The place we occupy seems all the world
>   (PMS, A61-47)

Clare can watch other living things lose their way, while feeling still sufficiently in control of his environment to write that 'The place we occupy seems all the world.' The balance of prospect-refuge symbolism is ideal: Clare can see out into the world, secure in his refuge of mist. The shepherds in this sonnet are thrown back on their inner life, on a solipsism foreign to the neo-classical writer. When Pope, in 'Windsor-Forest', describes a shepherd gazing at a reflection in a stream, he maintains a careful respect for distance, observer and observed are kept separate; the trees 'tremble in the Floods', but they are also specified as 'absent'. The couplets control the relationship between subject and object, as thesis and antithesis:

> Oft in her Glass the musing Shepherd spies
> The headlong Mountains and the downward Skies,
> The watry Landskip of the pendant Woods,
> And absent Trees that tremble in the Floods;
> In the clear azure Gleam the Flocks are seen,
> And floating Forests paint the Waves with Green.
> *(Pope: Poems* I. 169)

The best Romantic poems make a synthesis of those balanced antitheses (and the most banal attempt a synthesis where there is no thesis). When the Romantic looks at the reflection, he sees himself in the river of life, becomes Narcissus, occupied in his own space which seems 'all the world'. The whole landscape becomes a series of mirrors in which the poet reads aspects of himself.

Clare's vision is conditioned by pressures of immediate time and immediate space, and has much in common with that of primitive man. Pressures of time and space in the modern world have forced us into new perspectives, in which primitive art once again comes into its own:

Primitive and pre-alphabet people integrate time and space as one and live in an acoustic, horizonless, boundless, olfactory space, rather than in visual space. Their graphic presentation is like an X-ray. They put in everything they know, rather than only what they see. A drawing of a man hunting seal on an ice floe will show not only what is on top of the ice, but what lies underneath as well. The primitive artist twists and tilts the various possible visual aspects until they fully explain what he wishes to represent.[17]

In a similar fashion Clare 'twists and tilts' the 'visual aspects' of his world by giving the forces of nature a free hand:

> The shepherd rambling valleys white and wide
> With new sensations his old memorys fills
> When hedges left at night no more descried
> Are turned to one white sweep of curving hills
> And trees turned bushes half their bodys hide
>
> The boy that goes to fodder with supprise
> Walks oer the gate he opened yesternight
> The hedges all have vanished from his eyes
> Een some tree tops the sheep could reach to bite
> The novel scene emboldens new delight
> And though with cautious steps his sports begin
> He bolder shuffles the hugh* hills of snow
> Till down he drops and plunges to the chin
> And struggles much and oft escape to win
> Then turns and laughs but dare not further go
> For deep the grass and bushes lie below
> Where little birds that soon at eve went in
> With heads tucked in their wings now pine for day
> And little feel boys oer their heads can stray
>
> (*SPP* 143-4)

It is with a 'new delight' that is very much more than visual that Clare can write, 'The hedges all have vanished from his eyes'. The snow has removed at one stroke the omnipresent reminders of Clare's expulsion from Eden—the hedges of enclosure.

Rain can also cause strange perspectives and altered levels, a kind of elemental confusion:

> The maiden ran away to fetch the cloaths
> & threw her apron oer her cap & bows
> But the shower catchd her ere she hurried in
> & beat & almost dowsed her to the skin
> The ruts ran brooks as they would neer be dry
> & the boy waded as he hurried bye
> The half drowned ploughman waded to the knees
> & birds were almost drowned upon the trees
> The streets ran rivers till they floated oer
> & women screamed to meet it at the door
> Labour fled home & rivers hurried bye
> & still it fell as it would never stop

> Een the old stone pit deep as house is high
> Was brimming oer & floated oer the top
>
> (PMS, A61-79)

The element of hazard is more pronounced in this sonnet, but
the visual delight is just as keen. Rain, storms, high winds, and
snow necessitate shelter, and shelter is a form of enforced concealment, which can provide pleasure of the 'count your blessings' sort:

> How oft a summer shower hath started me
> To seek for shelter in a hollow tree
> Old hugh* ash dotterel* wasted to a shell
> Whose vigorous head still grew and flourished well
> Where ten might sit upon the battered floor
> And still look round discovering room for more
> And he who chose a hermit life to share
> Might have a door and make a cabin there
> They seemed so like a house that our desires
> Would call them so and make our gipsey fires
> And eat field dinners of the juicey peas
> Till we were wet and drabbled* to the knees
> But in our old tree house rain as it might
> Not one drop fell although it rained till night
>
> (SPP 162-3)

Again Clare is mimicking, this time with approval, the picturesque craze for a would-be hermit life, with its root-houses and
cottages so smothered in ivy that they look like trees. The hermit's delight in contrasted inner and outer conditions, what
Wordsworth calls 'audible seclusions', can be achieved even in
the roughest conditions:

> The firs that tapers into twigs & wear
> The rich blue green of summer all the year
> Softening the roughest tempests almost calm
> & offering shelter ever still & warm
> To the small path that ? levels underneath
> Where loudest winds almost as summers breath
> Scarce fans the weed that lingers green below
> When others as out of doors are lost in frost & snow

> & sweet the music trembles on the ear
> As the wind suthers through each tiney spear
> Make shifts for leaves & yet so rich they show
> Winter is almost summer where they grow
>
> (PMS, A57-1)

Clare's most splendid storm is in 'November' of *The Shepherd's Calendar*. In the following verse, he 'paints' a scene of 'irregularity'—a favourite word with the picturesque writers—but instead of a conventional canvas, it is as if Constable in his most inspired mood had dashed off a sketch, following Benjamin West's advice, 'always remember, sir, that light and shadow *never stand still*':[18]

> Dull for a time the slumbering weather flings
> Its murky prison round then winds wake loud
> Wi sudden start the once still forest sings
> Winters returning song cloud races cloud
> And the orison throws away its shrowd
> And sweeps its stretching circle from the eye
> Storm upon storm in quick succession crowd
> And oer the sameness of the purple skye
> Heaven paints its wild irregularity
>
> (*SC* 119)

To compare the first published version of 1827, 'improved' by John Taylor, with this 1964 version, edited from manuscripts by Robinson and Summerfield, is to compare a set piece prepared for an Academy exhibition in contrast to an original sketch; more presentable, perhaps, but less inspired:

> Yet but awhile the slumbering weather flings
> Its murky prison round—then winds wake loud;
> With sudden stir the startled forest sings
> Winter's returning song—cloud races cloud,
> And the horizon throws away its shroud,
> Sweeping a stretching circle from the eye;
> Storms upon storms in quick succession crowd,
> And o'er the sameness of the purple sky
> Heaven paints, with hurried hand, wild hues of every dye.[19]

The freshness and mobility of the original unpunctuated version
have been subordinated to the rules of those whom Clare called
'grammarian class-mongers' (PMS, A46-21). Most critics before
Robinson and Summerfield admired Clare through the well-
balanced panes of their study window. His response to nature
was too bracing for most tastes until well into this century:

> And quick it comes among the forest oaks
> Wi sobbing ebbs and uproar gathering high
> The scard hoarse raven on its cradle croaks
> And stock dove flocks in startld terrors flye
> While the blue hawk hangs oer them in the skye
> The shepherd happy when the day is done
> Hastes to his evening fire his cloaths to dry
> And forrester crouchd down the storm to shun
> Scarce hears amid the strife the poachers muttering gun
>
> The ploughman hears the sudden storm begin
> And hies for shelter from his naked toil
> Buttoning his doublet closer to his chin
> He speeds him hasty oer the elting* soil
> While clouds above him in wild fury boil
> And winds drive heavily the beating rain
> He turns his back to catch his breath awhile
> Then ekes* his speed and faces it again
> To seek the shepherds hut beside the rushy plain
> (*SC* 120-1)

Clare wrote *The Shepherd's Calendar* between 1821 and 1825 (it
was not published until 1827). By then he had learned, like
Turner, to 'describe' nature as 'fields of force' rather than
'separate objects in mechanical relations to one another'. Such
writing cannot be called 'descriptive' in any helpful sense.

Precise observation, too, is a necessary by-product of wary self-
preservation:

> A hugh* blue bird will often swim
> Along the wheat when skys grow dim
> Wi clouds—slow as the gales of spring
> In motion wi dark shadowd wing
> Beneath the coming storm it sails

And lonly chirps the wheat hid quails
That came to live wi spring again
And start when summer browns the grain
They start the young girls joys afloat
Wi 'wet my foot' its yearly note
So fancy doth the sound explain
And proves it oft a sign of rain

                                        (Ibid. 54)

Although Clare eschews didacticism, an essential component of georgic poetry, he is working here in the direct tradition of Virgil's First Georgic:

                                Never unawares
        Does rain attack men: either cranes descend
        From cloudland and take covert in deep vales,
        Or heifers sniff the breeze with nose in air,
        Or swallows circle shrieking round the pool,
        And mudlark frogs intone their ancient strain.
        Oft too the ant from inner sanctuaries
        Brings out her eggs, wearing a narrow path.
        The giant rainbow drinks, and from the fields,
        With humming of interminable wings,
        The rooks, an endless army, wend their way.[20]

In 'November', Clare echoes Virgil again, and lists the signs which foretell an autumn storm. In passages such as this, Clare has recaptured the vivid awareness of nature's beauty and treachery, found in the early pastoral poets:

        The hog sturts * round the stye and champs the straw
        And bolts about as if a dog was bye
        The steer will cease its gulping cud to chew
        And toss his head wi wild and startld eye
        At windshook straws—the geese will noise and flye
        Like wild ones to the pond—wi matted mane
        The cart horse squeals and kicks his partner nigh
        While leaning oer his fork the foddering swain
        The uproar marks around and dreams of wind and rain

                                        (SC 120)

Clare watches a landscape, rather than simply looking at it, which implies that he is actively involved in its processes, like an animal. He lets the reader share his probing of the hidden recesses of nature, a process which involves both constant motion and constant watchfulness. Jay Appleton writes, 'The activity of walking through a landscape is, in behavioural terms, the activity which introduces that "order" into a system of relationships which, without it, are haphazard and "chaotic".'[21] And E. H. Gombrich explains:

[Perception] is always an active process, conditioned by our expectations and adapted to situations. Instead of talking of seeing and knowing, we might do a little better to talk of seeing and noticing. We notice only when we look *for* something, and we look when our attention is aroused by some disequilibrium, a difference between our expectation and the incoming message.[22]

Again and again Clare reveals a naturalist's awareness of the significance of movement, no matter how slight, of the working of what Wordsworth calls the 'calm oblivious tendencies/ Of nature':

on the cow pasture I have often seen an hungry ox sturt * its head on one side making a snufting * noise and cease eating for a minute or two and then turn in another direction and on going to see what it turned from I have started up the old bird and found its nest often by this sign (*SPP* 108-9)

Clare inherited from the eighteenth century, of course, a solid tradition of perambulation. Whatever Dr Johnson thought about Spain, it is clear that there was not much of England that had 'not been perambulated' by Clare's youth. But Clare's walks are rarely interrupted for moralizing and never for the patriotic effusions found in Thomson and Cowper. His walks are random and informal in a way that the eighteenth-century poetry of perambulation seldom is, for the latter is usually connected with a moral, a train of thought, a plot, or simply a conversation. Clare's walks are always solitary, unlike Cowper's or Jane Austen's.[23]

When the eighteenth-century poet turns to prose, he can be more playful. Cowper and Gray often use the letter for their informal or frivolous moods. *The Task* has many exquisite touches, but nothing as delightful as the *fête-champêtre* with its 'wheel-barrow full of eatables and drinkables' described to the Revd William Unwin.[24] And Cowper, bubbling over to Lady Hesketh, forgets the *paysage moralisé* in his delight, 'I am daily finding out fresh scenes and walks, which you would never be satisfied with enjoying' (ibid., p. 172). These 'scenes and walks' are no less fresh when they get into verse, but they have to keep their place amidst passages of prolonged and serious meditation. Clare's poetry is 'daily finding out fresh scenes and walks', and there is little to prevent our direct enjoyment of them:

> Accross the fallow clods at early morn
> I took a random track where scant and spare
> The grass and nibbled leaves all closely shorn
> Leaves a burnt flat all bleaching brown and bare
>   (*SPP* 82)

In this chapter we have seen that, far from beginning anywhere and stopping anywhere, Clare crystallizes his descriptions into whole experiences. He does this by sensuous vitality, by co-operation with the forces of nature, and by keeping himself moving through the landscape. We have seen how Clare's relationship with external phenomena is a continuously active one, a process heightened by his habit of walking. By such means, Clare brings his verbal pictures alive. The next chapter examines his knowledge of art, and how he rejects the complex perspective of the Claudean tradition in painting in favour of the 'instantaneous sketches' of contemporary English artists, notably Peter De Wint.

4

'Natures Wild Eden'

Before the founding of the National Gallery in 1824, firsthand knowledge of great painting could only be obtained, apart from such bodies as the Royal Academy and the British Institution, by access to the numerous private collections up and down the country. Even today the pattern persists, and some of the finest Claudes, Stubbses, and water-colours are in private collections, brought out occasionally for large exhibitions. Horace Walpole could talk intimately about pictures because he was on intimate terms with their owners. Today the enthusiast must not only visit such places as the National and Tate Galleries, but also such houses as Althorp, Wilton, Woburn, Chatsworth, Petworth, and Burghley. If Clare had had the opportunity and the mobility, his knowledge might have been extensive, for it is clear that he was fascinated by pictures and by artists. But the odds against someone of his class obtaining firsthand knowledge were high. Even so, his knowledge was not negligible, and he took every opportunity to improve it. In his 'Autobiography', he tells us that on his second visit to London he 'did not know the way to any place for a long while but the royal academy & here I usd to go almost every day as Rippingille the painter had told the ticket keeper who I was & he let me come in whenever I chose which I often made use of' (*Prose*, p. 82).[1] Clare tells how Rippingille introduced him to Sir Thomas Lawrence (ibid., pp. 98-9). Clare also met H. F. Cary, the translator of Dante who lived in Hogarth's old house, where in the garden an inscription commemorated a dog, 'Life to the last enjoyed here Pompey lies', which Clare worked into a poem (see pp. 65-6). Clare ends the anecdote about Cary with a reference to the *ut pictura poesis* theory:

the Translator of Dante will not diminish the classical memorys of the old mansion with his possession of it Poetry & Painting are sisters (Ibid. 91)

Clare is not just repeating a cliché here to air his knowledge. He loved the sister art of painting and believed that Rippingille was doing with paint what he was attempting to do with words—to transcribe the freshness, secrecy, and wildness of nature without imposing a design or theory on it. For him 'Rippingille is the Theocritus of English painting' (ibid., p. 212).

Edward Villiers Rippingille (1798-1859) was one of a group of artists living in Bristol who were accustomed to regular sketching parties in the woods and glens around the city in search of idyllic landscapes.[2] He was a friend of Francis Danby (1793-1861), whose early landscapes are the group's most important legacy.[3] Rippingille's reputation sank very low after his death and he remains a relatively obscure figure, much of his early work, which was his best, having disappeared. He is chiefly remembered for his description of Turner on varnishing day in *The Art Journal* (1860). Clare describes him as 'a rattling sort of odd fellow' who 'has not been puffed into notice like thousands of farthing rush lights (like myself perhaps) in all professions that have glimmered their day & are dead ...',[4] and commends his 'true English conception of real pastoral life and reality of English manners and English beauty ...' (ibid., p. 172). Rippingille was also a native of the Fens, and Clare remembers having seen a shop window full of his work on a visit to Wisbech in 1809.[5] Clare was attracted to his conviviality and generosity but also to his love of children and his reverence for childhood. Danby writes of him, 'his *memory being excellent* and his feelings strong he recollected *almost with rapture* many of the little incidents that occurred to him in his boyhood.' (*The Bristol School of Artists*, p. 122.) This quality was likely to recommend itself to the poet who wrote, 'There is nothing but poetry about the existance of childhood ... and there is nothing of poetry about manhood but the reflection and the remembrance of what has been' (*SPP*, p. 18).

Rippingille made several unsuccessful attempts between 1822 and 1824 to entice Clare to Bristol. In one letter, as Eric Adams suggests, Rippingille may have been counting on Clare's memory of Danby's picture 'Clearing Up after a Shower', which Clare must have seen during his visits to the Royal Academy in 1822, especially if he was in Rippingille's company. As Adams writes, 'The landscape effect ... is pure Danby' (*Francis Danby*, p. 40):

Would you could have seen the sight I saw last Sunday and may see any day for the low charge of a three miles walk ... bunches of the most luxuriant foliage hanging over a river bright as a diamond forming the darkest recesses and hollows of the richest colours, the brown earth mixing with the endless varieties of greens produced from reflections and from the sun shining thro' broad sheets of leaves: see crowds of little brats dabbling in the sparkling water and their rosy feet, shining and paddling about in the mirror of all these beautiful objects around them, their vacant joyful faces, their eyes, their hair, their hearty laugh open and free as the blue vault over their heads.[6]

'Clearing Up after a Shower' (now lost), may itself have been inspired by one or more of Clare's own poems. Adams conjectures as follows: in November 1820 Clare's friend Mrs Emmerson presented her friends in Bristol, who included Rippingille, with a copy of Clare's first volume, *Poems Descriptive of Rural Life and Scenery*. She later sent Rippingille a copy of Clare's 1821 volume, *The Village Minstrel*, which contains a poem called 'Recollections after a Ramble', in which the poet shelters from a heavy shower, and then watches regretfully a group of boys playing with the rain-water in the brilliant sunshine which follows the shower (see *Poems* I. 181-8). Adams concludes, 'All we are left to suppose is the near-certainty that he [Rippingille] passed on his new enthusiasm to Danby, that Rippingille was responsible for Danby's becoming infected with the poetry of Clare, and Clare for Danby's attempting a picture in the manner of Rippingille.' (*Francis Danby*, p. 42.)

Another picture by Danby with a distinct flavour of Clare is 'Boys Sailing a Little Boat' (*c*.1821, City Art Gallery, Bristol; *The Bristol School of Artists*, frontispiece, p. 2). The incidental charm and commonplace details of this picture, as in many of Clare's poems, could easily lead to sentimentality, but Danby avoids this by his skilful manipulation of light and the unifying emotional intensity of his observation.

But the painter with whom Clare felt most emotional kinship was Peter De Wint (1784-1849).[7] Clare had met De Wint in

London and De Wint supplied the frontispiece to *The Shepherd's Calendar* in 1827, a picture of a harvest field with a group of harvesters at lunch, which harmonizes admirably with the mood of the poem. De Wint, however, was a hard man of business in his dealings and was less than generous to Clare, who had written to him requesting

a bit of your genius to hang up in a frame in my Cottage by the side of Friend Hiltons beautiful drawing which he had the kindness to give me when first in London what I mean is one of those scraps which you consider nothing after having used them & that lye littering about your study for nothing would appear so valuable to me as one of those rough sketches taken in the fields that breathes with the living freshness of open air & sunshine where the blending & harmony of earth air & sky are in such a happy unison of greens & greys that a flat bit of scenery on a few inches of paper appear so many miles (*Letters* 239)

Clare never received his scrap of genius.

Between 1821 and 1824, Clare wrote sonnets to Charles Lamb, Bloomfield, Izaak Walton, and the following sonnet to De Wint:

> Dewint I would not flatter nor would I
> Pretend to critic skill in this thine art
> Yet in thy landscapes I can well descry
> Thy breathing hues as natures counterpart
> No painted freaks—no wild romantic sky
> No rocks nor mountains as the rich sublime
> Hath made thee famous but the sunny truth
> Of nature that doth mark thee for all time
> Found on our level pastures spots forsooth
> Where common skill sees nothing deemed divine
> Yet here a worshipper was found in thee
> Where thy young pencil worked such rich surprise
> That rushy flats befringed with willow tree
> Rival'd the beauties of italian skies
>
> (*MC* 404)

This has no particular merit as a sonnet, with its stilted formalism and padding ('forsooth' for the rhyme, the eighteenth-century word 'pencil' for brush). But it is clear that Clare appreciates what makes De Wint's work distinctive and how it differs from a landscape in the Claude or Salvator Rosa tradition. The 'rich sublime' of Salvator is accurately delineated in the fifth and sixth lines, 'painted freaks', 'wild romantic sky', 'rocks', 'mountains'. The last line of the sonnet is probably a deliberate echo of Cowper's line, 'And throws Italian light on English walls'. Cowper is pleading for a more direct, sensuous approach to nature:

> Strange! there should be found,
> Who, self-imprison'd in their proud saloons,
> Renounce the odours of the open field
> For the unscented fictions of the loom;
> Who, satisfied with only pencil'd scenes,
> Prefer to the performance of a God
> Th' inferior wonders of an artist's hand!
> Lovely indeed the mimic works of art;
> But Nature's works far lovelier. I admire—
> None more admires—the painter's magic skill,
> Who shows me that which I shall never see,
> Conveys a distant country into mine,
> And throws Italian light on English walls:
> But imitative strokes can do no more
> Than please the eye—sweet Nature ev'ry sense.
>
> (*Cowper: PW* 138)

English painters of the Romantic school were soon to be busy throwing Dutch light on English walls,[8] and none more than De Wint, an Englishman of Dutch extraction. In the passage above, Cowper makes a firm distinction between 'the odours of the open field' and 'mimic works of art'. He writes that the 'imitative strokes' of art only 'please the eye', whereas nature involves 'ev'ry sense'. But for Clare, De Wint rivals nature in his involvement of the senses. His pictures are nature's 'autographs', not only seen but 'felt upon paper':

The only artist that produces real English scenery in which British landscapes are seen and felt upon paper with all their poetry & exhillerating expression of beauty about them is De Wint ... admirers of nature will admire his paintings—for they are her autographs & not a painters studys from the antique ... tis summer the very air breathes hot in ones face we see nothing but natural objects not placed for effect or set off by other dictates of the painters fancys but there they are just as nature placed them (*Prose* 212)

Clare's praise of De Wint as a kind of handmaid of nature can be
compared with Uvedale Price's praise of Claude, 'Every person
of the least observation must have remarked how broad the lights
and shadows are on a fine evening in nature, or (what is almost
the same thing), in a picture of Claude' (*Essays* I. 148). But Price
and Clare are using the word 'nature' here in quite different
senses. Clare knew that one crucial difference between these conceptions of nature is their use of space. The opening of Clare's
'Essay on Landscape' contains a juxtaposition of two kinds of
space, the neo-classical and space as in the paintings of De Wint.
The whole essay is an apologia for space as De Wint used it in
painting and as he himself used it in poetry:

There is no worse trickery of disposal of lights & shadows to catch the
eye from object to object with excessive fractions of diminishings untill
the eye rest upon that last pinspoint effect that makes a tree appear a
mile high & the neighbouring background a mile off—these beautiful
extravagances of false effects have produced such beautiful ugliness that
many have clomb into fame & profit by their creations but Dewint has
none of these minute gradations these atom stipplings by which beautiful effective compositions are produced but not paintings from Nature
as they profess to be (*Prose* 211)

Clare could well have had Claude in mind, and his description of
how the eye is led 'from object to object with excessive fractions
of diminishings' is an accurate analysis of what happens when
one looks at a Claude landscape. Compare two modern critics'
account of the Claude tradition:

... an artificial scheme, with a tall tree to one side in the foreground, a
'balancing' motif (trees or a building) situated further back on the other
side, and, at calculated intervals, a group of figures and a road or stream
leading the eye by stages to a range of hills on the horizon.[9]

For Clare, this is 'trickery'; he knew enough about painting and
about his own ambitions to be able to reject it.

But Clare had experience of another kind of formalized land-
scape which touched his life more closely—the landscape garden.
He was a gardener's assistant for a few months at Burghley
House, where over a period of twenty-five years Capability

Brown had completed one of his most impressive works. As well
as improving the park and creating the lake, Brown designed a
stable block, a greenhouse, a gamekeeper's lodge, a dairy with an
elaborate plaster ceiling, and a pinnacled summer-house, all in
the Gothick style. Clare must have seen these buildings.[10] He
tells us in his 'Autobiography' that he walked through the long
corridors of Burghley in his 'hard-nailed shoes' on his way to and
from an interview with the Marquess of Exeter, who became his
patron (*Prose*, pp. 68-9). On this occasion, presumably, he had
little time or desire to notice pictures, but it is likely that, as a
gardener's assistant, he would have been entertained in the kit-
chens and servants' quarters of the house and would have seen
something of the vast collection of pictures. Today there are
some 700 Old Master paintings, most of them collected before
the eighteenth century; a huge canvas of an ox, attributed to
Rubens, hangs in the old kitchen.[11]

Clare often walked through the park at Milton (seat of Earl
Fitzwilliam, another of his patrons) to visit his friends Artis and
Henderson, who were on the domestic staff and with whom he
exchanged botanical specimens. Humphry Repton had improved
this park in 1791.

Both Brown and Repton came under attack from the two main
protagonists of picturesque theory, Richard Payne Knight and
Uvedale Price. Although Payne Knight admired Claude, his
attacks on Brown in his poem *The Landscape* (1794) contain
sentiments close to Clare's:

> See yon fantastic band,
> With charts, pedometers, and rules in hand,
> Advance triumphant, and alike lay waste
> The forms of nature, and the works of taste!
> T'improve, adorn, and polish, they profess;
> But shave the goddess, whom they come to dress;
> Level each broken bank and shaggy mound,
> And fashion all to one unvaried round;
> One even round, that ever gently flows,
> Nor forms abrupt, nor broken colours knows;
> But, wrapt all o'er in everlasting green,
> Makes one dull, vapid, smooth, and tranquil scene.[12]

It is interesting to compare Knight's diatribe with the eulogy contained in an anonymous poem of 1767 in which Brown is compared to the great masters:

> At Blenheim, Croome and Caversham we trace
> Salvator's wildness, Claude's enlivening grace,
> Cascades and Lakes as fine as Risdale drew
> While Nature's vary'd in each charming view.
> To paint his works wou'd Poussin's Powers require,
> Milton's sublimity and Dryden's fire.
> Born to Grace Nature, and her works complete
> With all that's beautiful, sublime and great!
> For him each Muse enwreathes the Laurel Crown,
> And consecrates to Fame immortal Brown.[13]

This rather immodest flattery credits Brown with the qualities of the 'beautiful, sublime and great'. Although the poet here talks of Brown as if he were a great painter, he does not use the word 'picturesque'. It was not until the 1790s that the Picturesque was formulated as a theory to supplement the Sublime and Beautiful. Payne Knight and Price saw Brown as destroying true picturesque qualities, where 'forms abrupt' and 'broken colours' abound.

Repton was not a slavish imitator of Brown's methods (he stressed utility as much as beauty), but he is usually attacked or praised in the same breath as Brown. Thomas Love Peacock's novel *Headlong Hall* (1816) contains an amusing but sharp criticism of Repton (and by implication, Brown). Sir Patrick O'Prism is replying to Marmaduke Milestone (Repton):

Your system of levelling, and trimming, and clipping, and docking, and clumping, and polishing, and cropping, and shaving, destroys all the beautiful intricacies of natural luxuriance, and all the graduated harmonies of light and shade, melting into one another, as you see them on that rock over yonder. ... your improved places, as you call them ... are nothing but big bowling-greens, like sheets of green paper, with a parcel of round clumps scattered over them like so many spots of ink, flicked at random out of a pen, and a solitary animal here and there looking as if it were lost ...[14]

The tone here, under the banter, is that of Wordsworth's *Guide to the Lakes* and Clare's prose. Peacock's description of Brown's parks, 'like sheets of green paper', is especially apt, as Brown compared his art to punctuation. Brown's art was always deliberate, however, and his clumps punctuated his syntactical landscape in a studied way, not 'flicked at random out of a pen'. Nothing could better demonstrate the gulf between eighteenth- and nineteenth-century conceptions of nature than the fact that when Brown died (1783), an obituarist wrote of him, 'when he was the happiest man, he will be least remembered; so closely did he copy nature that his works will be mistaken'.[15] The nature that Brown copied is a far cry from 'Natures wild Eden wood and field and heath' (*SPP*, p. 115) which Clare loved and also attempted to transcribe faithfully.

One can understand Clare's feelings as he walked through Burghley or Milton. He wanted to experience nature in its primal naked freshness, whereas humans were always adorning this goddess with gauds and dressing her up for show. In one of his Red Books, Repton explains that 'where man resides, Nature must be conquered by Art, and it is only the ostentation of her triumph, and not her victory, that ought never to offend the correct Eye of Taste.' Repton's aim is 'la Simplicité soignée'.[16]

The metaphor of 'dressing' nature is a commonplace of the neo-classical tradition, epitomized by Pope's lines, '*True Wit* is *Nature* to Advantage drest/,What oft was *Thought*, but ne'er so well *Exprest*' (*Pope: Poems* I. 272-3).[17] Reynolds's *Discourses on Art* reiterate Pope's sentiments in lively prose. Burke uses the metaphor powerfully in his *Reflections on the Revolution in France*, revealing his detestation of the 'bare, forked animal':

All the decent drapery of life is to be rudely torn off. All the superadded ideas, furnished from the wardrobe of a moral imagination, which the heart owns, and the understanding ratifies, as necessary to cover the defects of our naked shivering nature, and to raise it to dignity in our own estimation, are to be exploded as a ridiculous, absurd, and antiquated fashion. (Harmondsworth, 1969, p. 171)

Jane Austen plays with the metaphor ironically in *Mansfield Park*. Mr Rushworth is advised by Miss Bertram to employ Repton to improve his estate; this idea is endorsed by all the least attractive characters in the book, including the odious Mrs Norris, so that Repton becomes the symbol of a false fashion, called in to fit out the 'old state' of Sotherton in a 'modern dress'.[18]

Clare, like Fanny Price in *Mansfield Park*, distrusts those who want the landscape to appear in new and fashionable clothes. He wants to see nature in 'her every day dessabille'. He wrote to De Wint telling him that his work had given him the courage of his conviction that few painters had looked at nature in her naked beauty:

> now I think many Painters look upon nature as a Beau on his person and fancies her nothing unless in full dress—now nature to me is very different & appears best in her every day dessabille in fact she is a Lady that never needed Sunday or holiday cloaths tho most painters & poets also have & still do consider that she does need little touches of their fancies & vagaries to make her beautiful which I consider deformities tho I should have given up the point in fancying that they might be right & I wrong if I did not feel that your sketches I speak of illustrated my opinion (*Letters* 239)

Crabbe made a similar rejection of the 'Sunday or holiday cloaths' of poetry in *The Village*:

> Then shall I dare these real ills to hide,
> In tinsel trappings of poetic pride?
> (Crabbe: *The Village*, University Tutorial Press (1950), p. 2)

Clare also castigates those 'modern fancy Landscapes' in which 'cows horses & sheep are scened cooling themselves in a pool which is out of nature for sheep were never seen in that situation since Noahs flood unless forced in for they have a great aversion to water' (*Letters*, p. 214). Martin Hardie writes of De Wint that 'he never invented the smallest figure; his harvesters and haymakers are never artificially grouped as pictorial adjuncts to enliven his scene. They seem to belong to the fields, to have grown from the soil as surely as his trees' (*Water Colour Painting in Britain* II. 222). De Wint catches the essence of English landscape without needing the reference kit of the mythological or history painter. Unlike Keats, in Clare's view, we are not always looking under a rose-bush to find Venus; we can take the roses for what they are:

we see the most natural reflection of scenery crowded with groups of satyrs & fawns & naiads & dryads & a whole catalogue of the vampire unaccountables dancing about in ridiculous situations round modern fishponds & immence temples in the Grecian style but no longer than a goose quill—& mown pleasure grounds kept as smooth & as orderly as a turkey carpet ... the very clouds were not out of the reach of these patronizing deformers—so they are loaded with fiddling Gods & gossiping Goddesses (*Prose* 214)

Clare's search for the 'natural' in art and nature often has remarkable affinities with Constable's. Constable writes of a landscape by Boucher:

From cottages adorned with festoons of ivy, sparrow pots, etc. are seen issuing opera dancers with mops, brooms, milk pails, and guitars. ... The scenery is diversified with winding streams, broken bridges, and water wheels; hedge stakes dancing minuets and groves bowing and curtseying to each other; the whole leaving the mind in a state of bewilderment and confusion, from which laughter alone can relieve it.[19]

Clare did not have the Romantic love of mountains, and he dismisses the excessive cult of ruins as 'claptrap'; De Wint's works bask in the 'poetry of light & sunshine'. Clare himself uses the dress metaphor in the word 'draperys', which suggests that, even for him, the goddess of nature modestly concealed her nakedness:

There are no mountains lifting up the very plains with their extravagant altitudes no old ruins with their worn & mossy claptrap for effect but simple woods spreading their quiet draperys to the summer sky & undiversified plains bask in the poetry of light & sunshine so void of all trick & effect (*Prose* 212)

De Wint himself said, 'I am never so happy as when looking at Nature. Mine is a beautiful profession' (*Water Colour Painting in Britain* II. 211). He is essentially a painter of the open air. Ruskin wrote that 'De Wint makes me feel as if I were walking through the fields' (ibid., p. 221). Clare had said something similar in the 'Essay on Landscape': 'if we could possibly walk into the picture we fancy we might pursue the landscape beyond those mysterys (not bounds) assigned it so as we can in the fields' (*Prose*, p. 211). It was this quality of intimacy which Constable and the Romantic painters admired in Hobbema, Ruysdael, Cuyp, and other Dutch masters. Claude was also admired for his portrayal of space, as if one could step into the frame and walk to the horizon. But Clare cannot feel at home in ideal landscape. One important difference between Claude and the Dutch painters is one of tone. Claude makes little allowance for the vagaries of the climate; his idealization is meteorological as much as topographical. Michael Kitson has pointed out that in the whole of Claude's work there is only one picture in which rain makes an appearance (and a peripheral one at that).[20] The golden tone of a Claude painting is the 'light that never was, on sea or land'. The Dutch masters, on the other hand, delight in conveying the nuances of localized, ever-changing light. Kenneth Clark explains that in Jacob van Ruysdael 'light has a new character. It is no longer static and saturating as in Bellini. It is in continual movement. Clouds pile up in the sky, shadows sail across the plain.'[21] De Wint captured this shifting light and shadow in his East Anglian scenes. Thackeray wrote of him, 'He spent his life in one revel of sunshine ... Fuseli, who wanted an umbrella to look at Constable's showers, might have called for a pot of porter at seeing one of De Wint's hay makings' (*Water Colour Painting in Britain* II. 221). This tangible feeling of elemental sparkle and movement is what Clare loved. It is nature's 'breathing hues', 'her every day dessabille', what many critics up to the end of the eighteenth century and beyond abusively called 'Dutchification'.

Critics in this period were about equally divided into those who believed that the Dutch and Flemish schools of painters were a bad influence and those who believed they were beneficial. Scott praised Jane Austen as a writer whose work 'reminds us something of the merits of the Flemish school of painting. The subjects are not often elegant, and certainly never grand; but they are finished up to Nature, and with a precision which delights the reader.'[22] The poetry of Crabbe was a test case for critics. He was often compared with Dutch painters, sometimes favourably, sometimes unfavourably. An anonymous reviewer of 1820 writes that his 'descriptions have generally the merit of a Dutch accuracy: sometimes they are strikingly picturesque and even beautiful'.[23] Another anonymous reviewer of 1819 takes the opposite view, employs the metaphor of clothes, and quotes Burke in the process:

Our contemporary poets ... have in their few laboured passages a Dutch minuteness of detail; and, generally careless of the grandeur or refinement of their subject-matter, they are satisfied with dressing their boors in appropriate trowsers, and painting the signposts of their ale-houses with broadly contrasted colours. In a word, poetry has been called down from that exalted region in which it was the delight of a few cultivated minds, and is now lowered to the pitch of the meanest intellect, and made the food of the vulgar. Its *'decent drapery'* is torn away, and its *ideal beauty* is prostrate in the dust. (Ibid. 272)

Hazlitt in 1821 writes (approvingly), 'the adept in Dutch interiors, hovels, and pig-styes must find in such a writer as Crabbe a man after his own heart' (ibid., p. 301). In his *Essay on Prints* (1792), Gilpin, in describing Antonie Waterloo's influence on the Romantic approach to nature, writes, 'We saw nature with a Dutchman's eye' (p. 110), but the following passage is more typical of his often cavalier treatment of nature:

I did all I could to make people believe they were *general ideas*, or *illustrations*, or any thing, but, what they would have them to be, exact portraits; which I had neither time to make, nor opportunity, nor perhaps ability; for I am so attached to my picturesque rules, that if nature gets wrong, I cannot help putting her right.[24]

In Clare's sonnet 'Wood pictures in winter', the local and particular details are what could be called 'Dutchification'. Add to this his identification with the moorhen and the hare, his use of local dialect words, the way he focuses the whole sonnet around the word 'picturesque', and one gets a good example of the Clare style:

> The woodland swamps with mosses varified
> And bullrush forrests bowing by the side
> Of shagroot sallows that snug shelter make
> For the coy morehen in her bushy lake
> Into whose tide a little runnel weaves
> Such charms for silence through the choaking leaves

> A whimpling* melodys that but intrude
> As lullabys to ancient solitude
> The wood grass plats* which last year left behind
> Weaving their feathery lightness to the wind
> Look now as picturesque amid the scene
> As when the summer glossed their stems in green
> While hasty hare brunts* through the creepy gap
> Seeks the soft beds and squats in safetys lap
>
> (*SPP* 141-2)

In this sonnet Clare is seeing 'nature with a Dutchman's eye', but unlike Gilpin, he is not so attached to his picturesque rules, 'that if nature gets wrong', he 'cannot help putting her right'. What Clare finds 'picturesque' is not a view of scenery which he can rearrange in his mind, but the grasses 'Weaving their feathery lightness to the wind'. He sees the grasses not just from above, where it is difficult to distinguish different species, but from the moorhen's point of view; from there they have not only poetic but ecological significance. Having noticed the shaggy and abrupt intricacy of the grasses outlined against the sky, stirred by the breeze (instead of the expected, one-dimensional word 'waves', Clare writes 'weaves', which catches the criss-cross motion), he will probably check his observation when he gets home, or when he next has access to books in the library at Milton, or he will jot down a note to his friend Henderson, or he will read his favourite poetry for mention of such an observation. In fact, the intricate outline of the grass, the 'bullrush forrests', and the 'shagroot sallows' all satisfy Gilpin's requirements for the Picturesque—roughness of texture, abruptness and discontinuity of line, and broken edges. Clare's picturesque eye has led him into a world which is poetically unique and ecologically differentiated, and once more he is at odds with neo-classical beliefs. Johnson may enunciate that 'a blade of grass is always a blade of grass, whether in one country or another',[25] but Clare's emotion arises both from his close perception of the grass and from his scientific instinct that this grass *does* differ from other grasses in the next county, let alone from country to country.

Water-colour depends for its effect upon the reflection of light from the paper; it is a transparent medium. With his distrust of 'trickery', perspective, and elevation, Clare was attracted by the fluid transparency of water-colour. De Wint

... had immense skill in dragging his brush over his granular paper so as to utilise the grain for sparkles of light and variety of texture. ... In all his happiest work he had the art of keeping large passages wet, leaving light spaces here and there, and running in rich warm colour to give his darks; then letting it all dry out quite undisturbed. That is why his dark tones are so fresh and luminous, so different from darks produced by two or more washes of superimposed colour. He had the courage to leave spaces that many painters would have sought to complete by further work. (*Water Colour Painting in Britain* II. 219, 220)

This note on De Wint's technique is worth bearing in mind as one reads such a poem as 'Wood pictures in Summer', which is an attempt to catch the liquid chiaroscuro of a water-colour in words. To say that it is a verbal De Wint is to praise it highly in its own right. There is the same 'deep softness of delicious hues' and 'endless lights and shades', the same feeling for moisture and texture, the same 'courage to leave spaces':

> The one delicious green that now prevades *
> The woods and fields in endless lights and shades
> And that deep softness of delicious hues
> That overhead blends—softens—and subdues
> The eye to extacy and fills the mind
> With views and visions of enchanting kind
> While on the velvet down beneath the swail *
> I sit on mossy stulp * and broken rail
> Or lean oer crippled gate by hugh * old tree
> Broken by boys disporting there at swee *
> While sunshine spread from an exaustless sky
> Gives all things extacy as well as I
> And all wood-swaily places even they
> Are joys own tennants keeping holiday
>
> (*SPP* 129-30)

This is more than a picturesque imitation. It has Clare's characteristic sweetness, which is a strength, because he is never sentimental. His profound emotion is present like the white grain of the water-colourist's background, occasionally given a vivid emphasis, but always there, against which his animated pictures move and sparkle.

It has already been mentioned how much Clare would have loved Bewick's work. He would also have responded deeply to the water-colourists of the Norwich school, especially Crome and Cotman, but also John Thirtle, Robert Leman, Thomas Lound, Henry Bright, and John Middleton. The very titles of their pictures could be titles of Clare poems, with their love of local detail and weather, 'A showery day', 'Cottage scene', 'River scene with rainbow', 'The study of a burdock', 'A thatched barn with cattle by a pond', 'Harling Gate', 'Return of the flock', 'Silver birches', 'Old cottage with gate on right', and 'The blasted oaks'.[26] John Crome's 'Mousehold Heath, Norwich' (oil on canvas, c.? 1818-20) comes as close as any contemporary painting to Clare's vision. It combines an uninterrupted view over undulating heath land to a spacious horizon below a huge sky, with a foreground which in itself is a micro-panorama of thistles, weeds, burdocks, and grasses. There is little conventional 'trickery', only two figures on a mound at the right, and a few cattle very informally scattered about.[27]

As the scientific exploration of fact proceeded apace, water-colour proved to be an ideal medium to record, with speed and sensitivity, botanical and zoological discoveries. From the beginning, therefore, water-colour was associated with the empirical way of looking at things which was to supersede the neo-classical outlook. As Graham Reynolds writes, 'The increased use of watercolour, in its modern connotation, came at the same period as the re-animation of man's interest in the visible world.' (*A Concise History of Watercolours* (London, 1971), p. 10.) It was also a medium suited to capture movement, fleeting effects of light, and the fickle moods of a landscape. Constable brilliantly exploited the empirical aspects of the medium in his studies of clouds.

The sonnet 'Wood Pictures in Spring' is another lively sketch, this time emerging from a confession of failure:

> The rich brown umber hue the oaks unfold
> When springs young sunshine bathes their trunks in gold

So rich so beautiful so past the power
Of words to paint—my heart aches for the dower
The pencil gives to soften and infuse
This brown luxuriance of unfolding hues
This living lusious tinting woodlands give
Into a landscape that might breath and live
And this old gate that claps against the tree
The entrance of springs paradise should be
Yet paint itself with living nature fails
—The sunshine threading through these broken rails
In mellow shades—no pencil eer conveys
And mind alone feels fancies and pourtrays
                                        (*SPP* 127)

This poem is far from perfect. As a sonnet it is irregular, with its run-on rhyming couplets all the way through; Clare is still using the word 'pencil' for brush, and his grammar and spelling are eccentric. But it remains a good example of his mature style, with its painter's sensitivity to colour, texture, light, and shade. A critic, not convinced of Clare's skill, might say, 'What does this sketchy poem do to distinguish it from hundreds of rather ordinary pictures?' Fuseli, when keeper of the Royal Academy, referred to 'the last branch of uninteresting subjects, that kind of landscape which is entirely occupied with the tame delineation of a given spot' (*Landscape into Art*, p. 49). Fuseli said this before Wordsworth and Constable had vindicated the cause of landscape art as an independent genre. Its vindication was hard won, however; we tend to forget how much against the grain of their time these two great figures were. 'Take away that nasty green thing' said the Royal Academicians when they saw Constable's 'Willows by a Stream',[28] and some of Wordsworth's most original poems received little but mockery until late in his long life. If Wordsworth and Constable had to struggle all their lives for the acceptance of their vision, how much chance did Clare have of critical favour?

A loving attention to the minutiae of nature is one thing; 'the tame delineation of a given spot' is another. Clare, as we saw in Chapter One, is no facile topographer. Even the barest or the tritest of his poems can be transformed by his extraordinary capacity for love, a love Franciscan in its purity. This capacity allows him to describe commonplace things without being commonplace. It is this quality, in addition to visual subtlety, that one would point out in defending the sonnet 'Wood Pictures in Spring'. Furthermore, this poem, while admitting that painters and poets cannot match the artistry of nature, ends on a Wordsworthian note: where art can go no further, the mind, in its loving intercourse with nature, takes over.

In an early sonnet, Clare had addressed the winds as 'Painters of nature!' (*Poems* I. 123.) But the sonnet is too much a conventional elegiac piece in the manner of W. L. Bowles to allow nature to display her artistry. In 'Wood Pictures in Spring', it is as if Clare has entirely withdrawn his ego from the scene, so that the mind and nature can act upon one another.

If the sceptic is still unconvinced that Clare can rise above mere word-painting, let him be reminded of Leigh Hunt's experiment. In *The Keepsake* for 1828 he wrote:

Did anybody ever think of painting a picture in writing? I mean literally so, marking the localities as in a map.
The other evening I sat in a landscape that would have enchanted Cuyp.
Scene—a broken heath, with hills in the distance. The immediate picture stood thus, the top and the bottom of it being nearly on a level in the perspective:

> Trees in a sunset, at no great distance from the foreground.
> A group of cattle under them, party-coloured,
> principally red, standing on a small landing place;
> the Sun coming upon them through the trees.
> A rising ground   A rising ground
>   Broken ground
> with trees.   with trees.
> Another landing place, nearly on a level
> with the cows, the spectator sitting and looking at them.

The Sun came warm and serious on the glowing red of the cattle, as if recognising their evening hues; and every thing appeared full of that quiet spirit of consciousness, with which Nature seems rewarded at the close of its day labours.[29]

It is amusing to notice how Hunt carefully preserves punctuation within his 'picture'. It would be difficult to make it more lifeless, and the coy commentary hardly helps. It reminds us that words are often at a loss when required to be taken literally. The glory of words, especially in poetry, is to describe the invisible, or to create a halo of emotion around the visible, as Clare does. Above all, words are not labels to stick into the frames of one's prejudices; they are living forces for good or ill. It is because Clare understood these living forces so well that he avoids the tame and literal delineation of given spots, making landscapes 'that might breath and live' in a 'wild Eden'.

Both Clare and his beloved De Wint created works with an unmistakable signature. De Wint, when asked why he never signed his pictures, replied that his pictures were 'signed all over'.[30] Clare could have said the same thing.

'Rich Instincts Natural Taste'

The preceding four chapters of this book have analysed the emergence of Clare's distinctive style from eighteenth-century traditions. Chapter One discusses how he lowers his vision from a bird's-eye to a beetle's-eye view, thereby creating a micro-panorama instead of a prospect; Chapter Two shows how this attention to the microscopic inevitably merges into an interest in natural history, and how a fascination with prolific detail leads to an outlook quite different from that of the neo-classical poet. In Chapter Three I discuss the kinds of problems encountered by Clare in his desire to describe his landscape while avoiding the obvious pitfalls of descriptive poetry, pitfalls much emphasized by representative eighteenth-century figures such as Pope and Johnson. Clare, I maintain, finds a way out of these difficulties by utilizing the kinetic elements inherent in the picturesque vision, as itemized by Praz: 'dazzle and flicker of effect, ... rapid succession of colours, lights, and shades, ... roughness, sudden variation, irregularity'. Clare not only admits more detail into his work than most of his predecessors, he is also aware that those details are in constant mutation. Simonides, in one of the key statements of the *ut pictura poesis* theory, writes that 'Painting is mute poetry, and poetry a speaking picture';[1] Clare would have agreed, but so much is the transitional and the dynamic stressed in his work that we are forced to rewrite the equation as 'Painting is static poetry, and poetry a moving picture'. This obsession with moving pictures is analysed in relation to Clare's knowledge of landscape gardening and his fascination with the forces of nature. His intimate acquaintance with the work of one of the great English water-colourists, Peter De Wint, is discussed in Chapter Four, where I stress the fact that De Wint's superiority, in Clare's eyes, arises from his delineation of

nature in her naked loveliness and his avoidance of the traditional methods of dressing her to advantage. In this fifth and
final chapter I intend to define Clare's use of the word 'taste'
more precisely, and in the process, tie up the various themes
discussed so far.

It is difficult to mention the word 'taste' without calling up certain social implications. 'Good taste' too often means
'fashionable taste'. The language of eighteenth-century aesthetics
was especially fraught with such implications, and Clare was as
instinctively suspicious of this language as he was of the
topographical language specified in Chapter One. In the eighteenth century, as Hermione Lee puts it, 'It was felt that taste for
art, taste for literature, and taste for nature, were related, and
were all three suggestive of one's moral worth and of one's social
standing.'[2] Shaftesbury, Burke, Hume, Hugh Blair, and Archi-
bald Alison all stressed the connection between aesthetic awareness and virtue; sometimes the association between sensibility,
virtue, and social respectability was stated openly, as in the
following advice on the art of landscape gardening by Humphry
Repton:

... the man of good taste ... knows that the same principles which direct
taste in the polite arts direct the judgement in morality; in short, that
a knowledge of what is good, what is bad, and what is indifferent,
whether in actions, in manners, in language, in arts, or science, con-
stitutes the basis of good taste and marks the distinction between the
higher ranks of polished society and the inferior orders of mankind.[3]

We take it so much as a truism today that virtue, sensibility, or
genius can emerge as readily from the cottage as from the palace
that we forget how much we are post-Wordsworthians in this
regard. For it was above all Wordsworth who claimed for the ordinary man, the obscure shepherd, even the madman, a place in
poetry. In 1819, when *Peter Bell* was finally published—one of
Wordsworth's boldest experiments in seeing 'how far the
language of conversation in the middle and lower classes of society is adapted to poetic pleasure'—it was greeted by a chorus of
ridicule. Leigh Hunt referred to it as 'another didactic little hor-
ror', full of 'half-witted prejudices'.[4] If Wordsworth's 'liberal'

contemporaries, as late as 1819, found his depiction of low life hard to take, they obviously had not outgrown the cultural snobbery inherent in such an eighteenth-century writer as Hume:

Sentiments which are merely natural affect not the mind with any pleasure, and seem not worthy of our attention. The pleasantries of a waterman, the observations of a peasant, the ribaldry of a porter or hackney coachman, all of these are natural, and disagreeable. ... Nothing can please persons of taste, but nature drawn with all her graces and ornaments, *la belle nature* ...[5]

Clare's attitude to *la belle nature* is that she is tastelessly over-dressed, tricked out to catch the eye of her lover, but not his true heart.

Although Clare's ideal of taste is obviously far from that of the Man of Taste, or connoisseur of *la belle nature*, he feels almost equally remote from the bleary-eyed rustic ('clown') who may see nature but who cannot feel it:

a clown may say that he loves the morning but a man of taste feels it in a higher degree by bringing up in his mind that beautiful line of Thomsons 'The meek eyd morn appears mother of dews' The rustic sings beneath the evening moon but it brings no associations he knows nothing about Miltons description of it ... nor of Collins Ode to Evening
      the man of taste looks on the little Celandine in Spring & mutters in his mind some favourite lines from Wordsworths address to that flower he never sees the daisy without thinking of Burns & who sees the taller buttercup carpeting the closes in golden fringe without a remembrance of Chatterton's beautiful mention of it ... other flowers crowd my imagination with their poetic assosiations but I have no room for them the clown knows nothing of these pleasures he knows they are flowers & just turns an eye on them & plods bye therefore as I said before to look on nature with a poetic eye magnifies the pleasure she herself being the very essence & soul of Poesy (*Prose* 175-6)

In the poem 'Shadows of Taste', Clare is even more outspoken:

> While the low herd mere savages subdued
> With nought of feeling or of taste imbued
> Pass over sweetest scenes a carless eye
> As blank as midnight in its deepest dye
> (*SPP* 113)

Clare not only opposes the 'clown' to the poetical man, but also to the scientific one. He does not regard science as an enemy of the imagination:

> The man of science in discoverys moods
> Roams oer the furze clad heaths leaf buried woods
> And by the simple brook in rapture finds
> Treasures that wake the laugh of vulgar hinds
> Who see no further in his dark employs
> Then village childern seeking after toys
> Their clownish hearts and ever heedless eyes
> Find nought in nature they as wealth can prize
> With them self interest and the thoughts of gain
> Are natures beautys all beside are vain
> But he the man of science and of taste
> Sees wealth far richer in the worthless waste
> Where bits of lichen and a sprig of moss
> Will all the raptures of his mind engross
> And bright winged insects on the flowers of may
> Shine pearls too wealthy to be cast away
> His joys run riot mid each juicy blade
> Of grass where insects revel in the shade
>
> (Ibid. 114-15)

Wordsworth had written in the Preface to *Lyrical Ballads* that true feeling does not need to be excited by 'gross and violent stimulants'.[6] Clare writes in a manuscript fragment, 'Pleasure with the man of taste is an evergreen it blooms with him thro all seasons & is always with him to gratify the mind' (NMS, A42-131).

In his joyful pursuit of 'bright winged insects', 'the man of science and of taste' would have earned censure from the eighteenth-century humanist. As Paul Fussell writes, 'the Augustan conservative imagination delights to image the contemptible by recourse to insects.'[7] With its respect for the hierarchy of the Chain of Being, the Augustan imagination is always on the alert to satirize those who would make themselves less than human by breaking this order of things: 'The image most accessible for this purpose ... is that of the nasty or showy or unstable insect.' (Ibid., p. 233.) Timon in Pope's *Epistle to Burl-*

ington is reduced to 'A puny insect, shiv'ring at a breeze!' (*Pope: Poems* III. ii. 143.) Shenstone advises the entomologists to regain an appropriately humanistic sense of balance:

> Let Flavia's eyes more deeply warm,
>   Nor thus your hearts determine,
> To slight dame Nature's fairest form,
>   And sigh for Nature's vermin.[8]

Goldsmith, in *History of the Earth, and Animated Nature* (1774), refers to insects as 'hideous', 'odious', 'nauseous', and 'obnoxious'.[9] And once again, a central passage of Burke is driven by the same contempt:

> By this unprincipled facility of changing the state as often, and as much, and in as many ways as there are floating fancies or fashions, the whole chain and continuity of the commonwealth would be broken. No one generation could link with the other. Men would become little better than the flies of a summer.[10]

It is in the light of these quotations that we should return to Clare's 'man of science and of taste' in his rapturous appreciation of insects, who are 'pearls too wealthy to be cast away' and who 'revel in the shade'. The insects here are not only valuable, they are merry. Clare, like Blake in 'The Fly', has once more reversed the humanistic assumptions, as he does in the prose fragment 'House or Window Flies':

> These little indoor dwellers, in cottages and halls, were always entertaining to me; after dancing in the window all day from sunrise to sunset they would sip of the tea, drink of the beer, and eat of the sugar, and be welcome all the summer long. They look like things of mind or fairies, and seem pleased or dull as the weather permits. In many clean cottages & genteel houses, they are allowed every liberty to creep, fly, or do as they like: and seldom or ever do wrong. In fact they are the small or dwarfish portion of our own family, and so many fairy familiars that we know and treat as one of ourselves. (*Prose* 251)

One would give a lot to hear Dr Johnson's reaction to such a passage, if presented with it by an eager Boswell. Not only are flies referred to with affection (no imagination nourished on *King*

Lear would ever find flies 'entertaining'), but they are accepted by Clare as 'portion of our own family'. If the Romantic Clare delights to 'sigh for Nature's vermin', the Augustan Swift rejects such a habit with characteristic scorn:

> Lice from your Body suck their Food;
> But is a Louse your Flesh and Blood?
> Tho' born of human Filth and Sweat, it
> May well be said Man did beget it.
> But Maggots in your Nose and Chin,
> As well may claim you for their Kin.
>
> . . . .
>
> Let Folks in high, or holy Stations,
> Be proud of owning such Relations;
> Let Courtiers hug them in their Bosom,
> As if they were afraid to lose 'em:
> While I, with humble *Job*, had rather,
> Say to Corruption—*Thou'rt my Father*.
> For he that has so little Wit,
> To nourish Vermin, may be *bit*.[11]

What we have in this contrast is an example of the conflicting viewpoints of Ancient and Modern. Even Thomas Hardy, with the irony of his post-Darwinian vision, can refer to insects with something like affection. In 'An August Midnight', for example, a longlegs, a moth, and a dumbledore (cockchafer) fly in his open window and bang against his lamp:

> 'God's humblest, they!' I muse. Yet why?
> They know Earth-secrets that know not I.[12]

Clare, in his verse and prose, is continually compelled to investigate those 'Earth-secrets', as in the poem 'Insects':

> Thou tiney loiterer on the barleys beard
> And happy unit of a numerous herd
> Of playfellows the laughing summer brings
> Mocking the suns face in their glittering wings
> How merrily they creep and run and flye
> No kin they bear to labours drudgery
> Smoothing the velvet of the pale hedge rose
> And where they flye for dinner no one knows

> The dewdrops feed them not—they love the shine
> Of noon whose sun may bring them golden wine
>
> (*SPP* 84)

It must be admitted that, just as Constable and Wordsworth can become banal when they are over-literal, Clare can be trite when he falls below his best. The images 'laughing summer', 'the velvet of the pale hedge rose' and 'golden wine' are almost tediously commonplace. And when he goes on to compare the insects to 'princes ... In silken beds' and 'fairey folk in splendid masquerade', one feels that he is anticipating the wrong sort of Victorian nature poetry, what Joyce was to describe as 'poetry for ladies':[13]

> Then in the heath bells silken hood they flie
> And like to princes in their slumber lie
> From coming night and dropping dews and all
> In silken beds and roomy painted hall
> So happily they spend their summer day
> Now in the corn fields now the new mown hay
> One almost fancys that such happy things
> In coloured moods and richly burnished wings
> Are fairey folk in splendid masquerade
> Disguised through fear of mortal folk affraid
> Keeping their merry pranks a mystery still
> Lest glaring day should do their secrets ill
>
> (*SPP* 84)

But Clare is too persistent and too prolific for 'Earth-secrets' to evade him for long. In the following verses from 'Emmonsales Heath' he conveys a sense of the quiet blessings which nature can give to those who approach it in the right spirit, and here the insects are described by an all-embracing love which is in no danger of becoming sentimental:

> In thy wild garb of other times
> I find thee lingering still
> Furze oer each lazy summit climbs
> At natures easy will
> ....

The birds still find their summer shade
To build their nests agen
And the poor hare its rushy glade
To hide from savage men

Nature its family protects
In thy security
And blooms that love what man neglects
Find peaceful homes in thee
. . . . : . . . .
Ive often met with places rude
Nor failed their sweet to share
But passed an hour with solitude
And left my blessing there
. . . .
O there are spots amid thy bowers
Which nature loves to find
Where spring drops round her earliest flowers
Uncheckt by winters wind
. . . .
Joy nursed me in her happy moods
And all lifes little crowd
That haunt the waters fields and woods
Would sing their joys aloud

I thought how kind that mighty power
Must in his splendour be
Who spread around my boyish hour
Such gleams of harmony

Who did with joyous rapture fill
The low as well as high
And make the pismires* round the hill
Seem full as blest as I

(Ibid. 166-8)

In Chapter One, I describe how Clare adopts an 'insect view', enabling him to see nature in an original way. This viewpoint was feared by the neo-classical writer because of its threat to the dignity of man; Paul Fussell explains that Pope's rejection of the 'microscopic eye' was strengthened by his memory of the 'revolting' engravings of 'a sixteen-inch flea and a twenty-one-inch louse' in Robert Hooke's *Micrographia* (1665).[14] One suspects that if Clare had been acquainted with these engravings, he would not have written of insects as 'fairey folk in splendid masquerade'. But even if he had known of such things, he would have been convinced that a great deal of nature had been needlessly rejected; that nature, for all its cruelty, had an underlying sanity and wholeness. Nature, in other words, revealed an instinctive taste in its proceedings which the Man of Taste was often incapable of appreciating. We must now define this 'natural taste' as Clare understood it.

In the poem 'Shadows of Taste', quoted in Chapter One (pp. 37-8) in connection with molehills, Clare plays sixteen variations upon the word 'taste'. The opening passage is one of the most interesting, and can bear repetition here:

> Taste with as many hues doth hearts engage
> As leaves and flowers do upon natures page
> Not mind alone the instinctive mood declares
> But birds and flowers and insects are its heirs
> Taste is their joyous heritage and they
> All choose for joy in a peculiar way
> Birds own it in the various spots they chuse
> Some live content in low grass gemmed with dews
> The yellow hammer like a tasteful guest
> Neath picturesque green molehills makes a nest
> Where oft the shepherd with unlearned ken
> Finds strange eggs scribbled as with ink and pen
> He looks with wonder on the learned marks
> And calls them in his memory writing larks

In the first two lines, Clare is saying that the creative signatures on nature's page are as infinite as individual leaves and flowers. One remembers one of Constable's sayings, 'The world is wide; no two days are alike, nor even two hours; neither were there ever two leaves of a tree alike since the creation of the world; and the genuine productions of art, like those of nature, are all distinct from each other.'[15] In 'The Nightingales Nest', Clare's praise of nature's fecundity contains a remarkable verbal echo of Constable, 'Her joys are evergreen her world is wide' (*SPP*, p. 74).

In the next four lines of 'Shadows of Taste' (lines 3-6) Clare says that birds, flowers, and insects are heirs to an instinctive joy which is beyond the grasp of man's rationality. Furthermore, this instinct is always allied to an infallible artistry, as evidenced by the taste of the yellow-hammer in selecting the site for its nest and the 'learned marks' on its eggs (lines 7-14). This conviction that animals, birds, and flowers have what can almost be called an aesthetic dimension was shared by Wordsworth, 'And 'tis my faith that every flower / Enjoys the air it breathes.' (*Wordsworth:PW* IV.58). Even the level-headed Gilbert White uses the phrase 'these little artists' in discussing the nesting habits of the bank-martin:

It is curious to observe with what different degrees of architectonic skill Providence has endowed birds of the same genus, and so nearly correspondent in their general mode of life! ... the bank-martin terebrates a round and regular hole in the sand or earth, which is serpentine, horizontal, and about two feet deep. At the inner end of this burrow does this bird deposit, in a good degree of safety, her rude nest, consisting of fine grasses and feathers, usually goose feathers, very inartificially laid together. ... In what space of time these little artists are able to mine and finish these cavities I have never been able to discover. (*Selborne* 178-9)

To write about insects as 'fairey folk' is one thing; to describe insects and birds as 'little artists', as Clare and Gilbert White do, is quite another. The first phrase is a vague poeticism; the second would not be dismissed by the scientist, especially the modern entomologist and ornithologist. Here, once again, we must leave the confines of literary criticism, and return to Jay Appleton's thesis in his book *The Experience of Landscape*. Appleton, in discussing inborn and learned behaviour in animals, and the possibility that aesthetic awareness has atavistic origins, quotes John Dewey:

To grasp the sources of the esthetic experience it is, therefore, necessary to have recourse to animal life below the human scale. The activities of the fox, the dog, and the thrush may at least stand as reminders and symbols of that unity of experience which we so fractionize when work is labor, and thought withdraws us from the world.   (p. 59)

In other words:

> Not mind alone the instinctive mood declares
> But birds and flowers and insects are its heirs

Dewey continues:

I do not see any way of accounting for the multiplicity of experiences of this kind ... except on a basis that there are stirred into activity resonances of dispositions acquired in primitive relationships of the living being to its surroundings, and irrecoverable in distinct or intellectual consciousness. (*The Experience of Landscape* 59)

Judged by the standards of his society (or of any society), Clare's extreme retreat from the world of men to the world of nature was an act of escapism. But seen in the light of modern research, Clare was instinctively turning his back upon what Coleridge calls 'that inanimate cold world' to find 'The passion and the life' in even the lowest forms of existence. The goings-on of the natural world comprise an animated universe, driven by passion and only explicable in terms of such human concepts as joy and taste. Keats writes in one of his letters, 'I go among the Feilds and catch a glimpse of a stoat or a fieldmouse peeping out of the withered grass—the creature hath a purpose and its eyes are bright with it ...'.[16] Clare's dissatisfaction with Keats's need for sentiment in poetry has already been quoted, but the Keats of the letters (whom some critics consider a greater poet than the Keats of the poems) would have been more congenial to him. Was there ever a poet who embodied Keats's doctrine of Negative Capability more completely than Clare? Clare would have recognized a kindred spirit in Keats's humble empathy with nature and his strength of purpose, '... if Poetry comes not as naturally as the Leaves to a tree it had better not come at all', '... let us open our leaves like a flower and be passive and receptive', '... if a Sparrow come before my Window I take part in its existince and pick about the Gravel.' (Ibid., pp. 70, 66, 38.) Keats and Clare both demonstrate those 'primitive relationships of the living being to its surroundings' in their ability to 'pick about the Gravel', and are aware that what I have called an 'atavistic alertness' (p. 88) is 'irrecoverable in distinct or intellectual consciousness'. Keats, Coleridge, Clare, and Wordsworth are among those Romantic figures who believe that nature is fundamentally a totality and that psychic balance is lost when man becomes, in Wordsworth's phrase, 'An intellectual All-in-all!' (*Wordsworth: PW* IV.66.) This belief is expressed by the modern environmentalist, Harold F. Searles, as follows:

It is my conviction that there is within the human individual a sense *of relatedness to his total environment*, that this relatedness is one of the transcendentally important facts of human living, and that if he tries to ignore its importance to himself, he does so at peril to his psychological well being. (*The Experience of Landscape* 67)

The Romantic stress on the uniqueness of living things embraces not only the human individual but delights in differentiating the leaves of trees, the shapes of clouds, blades of grass, and the artistry and 'architectonic skill' of birds and animals. This stress on subtle, often minuscule distinctions, is of course one of the hallmarks of the scientist. Appleton quotes the work of W. H. Thorpe on chaffinches to illustrate his belief that animal behaviour is both inborn and learned:

By using oscillograms in the study of the songs of chaffinches Thorpe was able to demonstrate that ' ... individual differences are not the expression of genetic differences but develop by learning during the early life of the bird.'

In these experiments it was shown that a young chaffinch, reared separately out of hearing of all chaffinch song, developed a very restricted song devoid of the usual phrasing and of the elaborate final flourish. 'The simple, restricted song of the isolated bird can be taken to represent the inherited basis of a chaffinch's performance.... In nature young chaffinches must certainly learn some details of song from their parents or from other adults in the first few weeks of life. ... But not until the critical learning period, during the following spring, does the bird develop the finer details of its song. This is the time when the young wild chaffinch first sings in a territory in competition with neighboring birds of the same species, and there is good evidence that it learns details of song from these neighbors. ... So the full chaffinch song is a simple integration of inborn and learned song patterns, the former constituting the basis for the latter.'

Appleton concludes, 'The individual creature interacts with its environment in a manner which, in the most general sense, is common to its species, but in detail is peculiar to itself.' (Ibid., pp. 59-61.) Towards the end of his book, Appleton produces a definition of taste which he believes can satisfy widely divergent viewpoints, '*Taste is an acquired preference for particular methods of satisfying inborn desires.*' (Ibid., p. 237.)

In the light of these quotations from Thorpe and Appleton, one can reconsider Daines Barrington's description of the nightingale's song, quoted on p. 51:

> ... I have observed sixteen different beginnings and closes, at the same time that the intermediate notes were commonly varied in their succession with such judgment as to produce a most pleasing variety.... Whenever respiration, however, became necessary, it was taken with as much judgment as by an opera singer.... the bird also sings (if I may so express myself) with superior judgment and taste.

Barrington's use of the words 'judgment' and 'taste', far from being poetic licence, turns out to have a basic scientific accuracy. Birds are not wound-up toys and are not mechanically predictable except 'in the most general sense'. Barrington is not being fanciful in describing the 'sixteen different beginnings and closes' and 'pleasing variety' of the nightingale's song as showing 'judgment' and 'taste', for these constitute that part of its experience which 'in detail is peculiar to itself'; it is learned behaviour, or 'an acquired preference for particular methods of satisfying inborn desires'.

Similarly, Clare's description of the nightingale's song, quoted on pp. 52 and 53, is an accurate appreciation of the bird's acquired preferences. The traditional literary description of the nightingale's song is 'jug jug jug'; Clare adds to this the distinctive grace-notes of the individual bird: 'Chew-chew Chew-chew' ... 'Cheer-cheer Cheer-cheer' ... 'Cheer-up Cheer-up cheer-up' ... 'Tweet tweet tweet jug jug jug' ... 'Wew-wew wew-wew chur-chur chur-chur 'Woo-it woo-it' ... 'Tee-rew Tee-rew tee-rew tee-rew 'Chew-rit chew-rit' ... 'Will-will will-will grig-grig grig-grig' ... 'tweet tweet tweet' ... 'jug jug jug'.

In 'The Moorehens Nest' (from which I quote on p. 47), the

birds are valued for their taste as picturesque architects; the selection of the site is almost a problem in aesthetics:

> Though less beloved for singing then the taste
> They have to choose such homes upon the waste
> Rich architects—and then the spots to see
> How picturesque their dwelling makes them be
> The wild romances of the poets mind
> No sweeter pictures for their tales can find

The sand-martin (see p. 54) is praised as he proceeds 'With strangest taste and labour undeterred / Drilling small holes along the quarrys side'.

The mole, as we saw in Chapter One (p. 38), is also praised as a 'Rude architect' who is heir to 'rich instincts natural taste'. We can now see that what Clare means here is that the mole has certain inborn capacities or instincts which he learns to adapt to each unique situation, thereby acquiring 'natural taste'. The taste (or 'acquired preference') is natural as opposed to artificial (in its pejorative sense) because it is quite simply a matter of survival for the mole to be a skilful landscape architect:

> No rude inellegance thy work confounds
> But scenes of picturesque & beautiful
> Lye mid thy little hills of cushioned thyme

Gilpin writes in his *Three Essays*, 'There are few parts of nature, which do not yield a picturesque eye some amusement' and continues in a vein which Clare would have appreciated, 'The more refined our taste grows from the *study of nature,* the more insipid are the *works of art.* ... the varieties of nature's charts are such, that, study them as we can, new varieties will always arise: and let our taste be ever so refined, her works, on which it is formed ... must always go beyond it ...'.[17] But Gilpin can never leave it at that; if at one moment he is full of enthusiasm, at the next the schoolmaster takes over, 'We must ever recollect that nature is most defective in composition; and *must* be a little assisted. Her ideas are too vast for picturesque use, without the restraint of rules. ... under these circumstances we see nature in her best attire, in which it is our business to

describe her.' (Ibid., pp. 67, 75.) And in his poem 'On Landscape
Painting', which appears at the end of the *Three Essays*, Gilpin
writes:

> We never mean,
> With close and microscopic eye, to pore
> On every studied *part*.
> . . . .
> Artists there are,
> Who, with exactness painful to behold,
> Labour each leaf, and each minuter moss,
> Till with enamelled surface all appears
> Compleatly smooth. Others with bolder hand,
> By Genius guided, mark the general form,
> The leading features, which the eye of taste,
> Practised in Nature, readily translates.
>
> (Ibid. 118, 117)

The difference between Clare's 'natural taste' and Gilpin's 'eye of
taste, Practised in Nature' is pin-pointed in the verb 'translates'.
For all his love and appreciation of nature, Gilpin still regards
natural beauty as written in a kind of foreign language, which
must then be translated into human terms. But Clare sees it as
his duty to learn that language, not to translate it or to dress it up
'in her best attire', guided by 'Genius'. This act of translation
reminds him all too forcibly of his own 'translation' from rapturous boyhood in the 'ancient passion' of an open field landscape
to disenchanted manhood in the 'foreign land' of enclosure:

> But take these several beings from their homes
> Each beautious thing a withered thought becomes
> Association fades and like a dream
> They are but shadows of the things they seem
> Torn from their homes and happiness they stand
> The poor dull captives of a foreign land
>
> (*SPP* 116)

The learning of this language requires a humility which Genius
is apt to overlook. Clare explores the mole's territory as a 'guest'
and suggests in the last line of his sonnet that true taste in art is

achieved by watching such skilled labourers as the mole:

> ... when I climb
> Thy little fragrant mounds I feel thy guest
> & hail neglect thy patron who contrives
> Waste spots for the[e] on natures quiet breast
> & taste loves best where thy still labour thrives

This identification with nature is a dangerous process, of course. Nature is nature and art is art, and sometimes Clare seems to blur the distinction between them. Complete nakedness of spirit, like undefiled innocence, cannot be achieved in this world. In his desire to be 'too like nature' (see p. 39), Clare is, in a sense, swallowed up by nature; there is an Asylum poem in which he describes a bee dying 'self poisoned in a treacherous flower'.[18] The landscape which he saw with such acuity eventually became, in his madness, a living tomb. Once again, one is reminded of Richard Dadd's 'The Fairy Feller's Masterstroke', and the hopeless faces of the human insects who inhabit their dreadful miniature Arcadia. They have disdained the 'bounds of place and time' and are adrift on a 'living sea of waking dreams':

> I fled to solitudes from passions dream
> But strife persued—I only know I am.
> I was a being created in the race
> Of men disdaining bounds of place and time—
> A spirit that could travel o'er the space
> Of earth and heaven—like a thought sublime,
> Tracing creation, like my maker, free—
> A soul unshackled like eternity,
> Spurning earth's vain and soul debasing thrall
> But now I only know I am—that's all.
>
> (*SPP* 196)

But the causes of Clare's madness are beyond the scope of this book.

Taste for Clare, then, comprises what Price calls an 'active agency' (p. 82), what Middleton Murry calls an 'active relation' between subject and object (p. 77), the 'active process' of perception described by Gombrich (p. 95), the participation in landscape described by Appleton, in which the 'environment is translated from a passive to an active role' (p. 86), and a faculty which I have called 'dynamically selective' (p. 82). Clare believes that animals, birds, flowers, and insects have an inherent dynamism which is not reducible to the dictates of mechanical rules, but which operates from the laws of its own being, thereby demonstrating 'acquired preferences' when confronted by unique situations requiring creative action. Here, Clare is close to the critical theory of Coleridge, but he does not have Coleridge's faith in the power of the mind to dissolve, fuse, and reorder the minutiae of nature. He places his faith, perhaps too innocently, in the creative 'disorder' of nature, which becomes for him 'the truth of taste' as opposed to 'arts strong impulse'—the tyrannies of artifice:

> Some spruce and delicate ideas feed
> With them disorder is an ugly weed
> And wood and heath a wilderness of thorns
> Which gardeners shears nor fashions nor adorns
> No spots give pleasure so forlorn and bare
> But gravel walks would work rich wonders there
> With such wild natures beautys run to waste
> And arts strong impulse mars the truth of taste
>
> (*SPP* 116)

This passage towards the end of the poem 'Shadows of Taste' pushes almost to the limit that reaction against artifice and 'the formal Mockery of princely Gardens' initiated by Shaftesbury in the passage already quoted (p. 16), '... where neither Art, nor the Conceit or Caprice of Man has spoil'd their genuine order, by breaking in upon that primitive State.'

In his fear of formality, Clare certainly overemphasizes the 'primitive' qualities in nature; it has been the critical aim of this book to pin-point the 'genuine order' in his work. I believe that the sensitive reader cannot deny that the artistry of Clare's best work is of a high order. What will be contested, perhaps indefinitely, is Clare's right to a place beside the other six major Romantics. Those who prize above all intellectual complexity in poetry will always deny him high status. But Clare, I claim, does have a certain kind of greatness, an unsensational resilience

which more volatile and brilliant spirits might well envy.
Through the harshest conditions, Clare maintained a unique vision with a toughness which kept his enormous capacity for love
alive:

> Poets love nature and themselves are love;
> The scorn of fools and mock of idle pride
> The vile in nature worthless deeds approve
> They court the vile and spurn all good beside
> Poets love nature like the calm of heaven
> Her gifts like heaven's love spread far and wide
> In all her works there are no signs of leaven
> Sorrow abashes from her simple pride
> Her flowers like pleasures have their seasons birth
> And bloom through regions here below
> They are her very scriptures upon earth
> And teach us simple mirth where e'er we go
> Even in prison they can solace me
> For where they bloom God is, and I am free.
>
> (Ibid. 197-8)

Genuine poetry, for Clare, is a form of love. This belief is sus-
tained by an active imagination whose originality lies in its basic
sanity, and whose charm stems from an extraordinary freshness
of perception. Clare speaks to us with quiet insistence from a
depth of experience which is universal. Once we have learned to
listen, we can never forget that haunting music.

NOTES

Introduction

1. 'Poems of John Clare's Sanity', *Some British Romantics*, ed. John Jordan, James Logan, and Northrop Frye (Ohio, 1966), pp. 189-232.

2. See *The Visionary Company* (Ithaca, 1971), pp. 444-56.

3. *The Shield of Achilles* (New York, 1944), p. 27.

4. J. W. and Anne Tibble, eds., *The Letters of John Clare* (London, 1951), p. 133. Hereafter cited as *Letters*.

5. Dorothy Stroud, *Capability Brown* (London, 1975), p. 201.

6. W. H. Auden, ed., *Nineteenth Century Minor Poets* (London, 1967), pp. 17-18.

7. Clare's collection of some 440 books is preserved in the Northampton Public Library. It is particularly strong in poetry and natural history. See *Catalogue of the John Clare Collection in the Northampton Public Library* by David Powell, pp. 23-34.

8. *TLS*, 9 June 1972, p. 654.

9. Joseph Addison, 'An Essay on Virgil's *Georgics*', in *The Miscellaneous Works of Joseph Addison*, ed. A. C. Guthkelch (London, 1914), II. 4.

10. See E. D. H. Johnson, ed., *The Poetry of Earth* (New York, 1974), p. viii.

11. The Picturesque emerged as a third category when Burke's much-quoted categories of Sublime (limitless, awe-inspiring) and Beautiful (smooth, curvilinear) were found inadequate to cover a whole range of objects strictly neither sublime nor beautiful. Gilpin sought the qualities of great landscape paintings in uncultivated nature, Uvedale Price extended and refined Gilpin's values, stressing the complexity of pictorial design, and Payne Knight emphasized mental energies expressed in the work of art. See Martin Price, 'The Picturesque Moment', *From Sensibility to Romanticism*, ed. F. W. Hilles and Harold Bloom (New York, 1965), pp. 259-92; also W. J. Hipple, *The Beautiful, The Sublime and The Picturesque in Eighteenth-Century British Aesthetic Theory* (Carbondale, 1957), pp. 185-320.

12. Quoted by J. I. M. Stewart in the *TLS*, 18 April 1975, p. 415.

13. From 'The Farmer's Boy' by Robert Bloomfield. *The Works of Robert Bloomfield and Henry Kirke White* (London, 1871), p. 66.

14. The publication of this volume is an event of considerable importance for students of Clare. The collection, which contains much of Clare's mature work, was ready for publication in 1832, but Clare was unable to find a sufficient number of subscribers. The editors, Anne Tibble and R. K. R. Thornton, have followed as closely as possible Clare's original manuscript,

now in the Peterborough Museum. Of the 361 poems in the volume, almost a third are published for the first time.

15. For a discussion of the Tibble editorial policy see the review of J. W. and Anne Tibble's *John Clare: His Life and Poetry*, *TLS*, 27 April 1956, p. 252; and related correspondence, *TLS*, 11 May 1956, p. 283; *TLS*, 25 May 1956, p. 313; *TLS*, 18 October 1957, p. 625.

Chapter 1. 'Picturesque green molehills'

1. Anne Tibble and R. K. R. Thornton, eds., *John Clare: The Midsummer Cushion* (Northumberland, 1979), p. 292. Hereafter cited as *MC*.

2. See Leslie Parris, *Landscape in Britain: c. 1750-1850* (London, 1973), pp. 124-5.

3. Raymond Lister, *British Romantic Art* (London, 1973), p. 143.

4. R. R. Wark, ed., *Discourses on Art* (London, 1966), p. 222.

5. E. W. Manwaring, *Italian Landscape in Eighteenth Century England* (New York, 1925), p.186.

6. Paget Toynbee and Leonard Whibley, eds., *Correspondence of Thomas Gray* (Oxford, 1971), p. 1079.

7. J. W. and Anne Tibble, eds., *The Prose of John Clare* (London, 1951), p. 43. Hereafter cited as *Prose*.

8. John Butt, ed., *The Poems of Alexander Pope* (London and New Haven, 1939-61), I. 246. Hereafter cited as *Pope: Poems*.

9. Ernest de Selincourt, ed., *The Poetical Works of William Wordsworth* (Oxford, 1940-49), II. 262. Hereafter cited as *Wordsworth: PW*.

10. *The Birth and Rebirth of Pictorial Space* (London, 1967), p. 123.

11. T. H. Banks, ed., *The Poetical Works of Sir John Denham* (New Haven, 1969), p. 77.

12. H. M. Margoliouth, ed., *The Poems and Letters of Andrew Marvell* (Oxford, 1971), p. 43.

13. *The Moralists*, Vol. ii. See Nikolaus Pevsner, *Studies in Art, Architecture and Design* (London, 1968), I. 79-101.

14. The winter garden which Wordsworth created for Sir George Beaumont at Coleorton in Leicestershire still exists. The garden, sunk behind a huge wall (once a quarry), retains some magic. See 'How Wordsworth made a garden', *The Times*, 12 August 1960; and Russell Noyes, *Wordsworth and the Art of Landscape* (Bloomington, 1968), Chapter 3.

15. *Three Essays: on Picturesque Beauty; on Picturesque Travel; and on Sketching Landscape* (London, 1792), p. 26.

16. *Fragments on the Theory and Practice of Landscape Gardening* (London, 1816), p. 235. Quoted (with plates) in *Landscape in Britain*, p. 61.

17. Eric Robinson and Geoffrey Summerfield, eds., *Selected Poems and Prose of John Clare* (London, 1967), p. 170. Hereafter cited as *SPP*.

18. C. B. Tinker, *Nature's Simple Plan* (New Jersey, 1922), p. 27.

19. *Wordsworth and the Art of Landscape*, p. 52.

20. Ernest de Selincourt, ed., *The Prelude* (Oxford, 1959), p. 439.

21. *Picturesque Landscape and English Romantic Poetry*, pp. 79-87.

22. 'Biographia Literaria', Chapter XIII, *Coleridge: Select Poetry and Prose*, ed. Stephen Potter (London, 1950), p. 246.

23. See R. A. Aubin, *Topographical Poetry in XVIII-Century England* (New York, 1936), pp. 298-314.

24. *The Poetical Works of Sir John Denham*, p. 63.

25. *Topographical Poetry in XVIII-Century England*, p. vii. [Aubin's italics.]

26. 'A Redefinition of Topographical Poetry', *Journal of English and Germanic Philology*, LXIX (1970), 394-406.

27. The five structural devices which Foster examines are: the creation of three-dimensional space, the use of space as a patterning device, the use of time-projections, the use of extended metaphor, and the development of a controlling moral vision.

28. 'The Measure of Paradise: Topography in Eighteenth-Century Poetry', *Eighteenth-Century Studies*, IX (Winter 1975/6), 232-56.

29. 'The Topographical Tradition in Anglo-Irish Poetry', *Irish University Review*, IV (Autumn 1974), 175.

30. *The Idea of Landscape and the Sense of Place 1730-1840: an Approach to the Poetry of John Clare* (Cambridge, 1972), pp. 64-97. Hereafter cited as *The Idea of Landscape*. This book opened a new phase in Clare studies, and I am indebted to Barrell's discussion of several aspects of Clare's work.

31. J. W. Tibble, ed., *The Poems of John Clare* (London, 1935), II. 181-212. Hereafter cited as *Poems*.

32. Although the word 'kaleidoscopic' is employed in this chapter in its figurative sense, it is significant that the instrument itself was invented in 1817. According to the *OED* one of the earliest references in English to 'Telescopioes' is by Boyle in 1648, six years after the publication of 'Cooper's Hill'. In discussing Clare's rejection of an instrumental vision, I am aware that the kaleidoscope itself is an optical instrument. But the crucial difference is that kaleidoscopes were not part of surveyors' equipment, and therefore would not have been associated by Clare with enclosure, landscape gardening, and the railways.

33. Mark Storey, ed., *Clare: The Critical Heritage* (London, 1973), pp. 197-8, 190, 195. Hereafter cited as *Heritage*.

34. See Eric Robinson and Geoffrey Summerfield, 'John Taylor's editing of Clare's *The Shepherd's Calendar*', *The Review of English Studies*, NS, XIV (November 1963), 359-69.

35. W. J. Bate and A. B. Strauss, eds., *The Yale Edition of the Works of Samuel Johnson* (New Haven, 1969), III. 197.

36. Nikolaus Pevsner, ed., *The Picturesque Garden and its Influence outside the British Isles* (Washington, 1974), p. 63.

37. See the footnote to the 1793 text of 'Descriptive Sketches', *Wordsworth: PW*, I. 62.

38. *The Natural History of Selborne* (London, 1971), pp. 4-5. Hereafter cited as *Selborne*.

39. *Poems on Several Occasions* (London, 1736), pp. 207-8.

40. *A Muse in Livery* (London, 1732), pp. 19-20.

41. See Rayner Unwin, *The Rural Muse* (London, 1954) for an account of these poets.

42. This is the title as it appears in *Poems Descriptive of Rural Life and Scenery* (London, 1820), p. 65. The title in Clare's manuscript is 'Elegy/Hastily composed & Written with a Pencil/on the Spot/In the Ruins of Pickworth/ Rutland'.

43. See John Dixon Hunt, *The Figure in the Landscape: Poetry, Painting, and Gardening during the Eighteenth Century* (Baltimore, 1976), p. 201. This excellent book is indispensable reading for those concerned with eighteenth-century landscape aesthetics.

44. See John Barrell, *The Idea of Landscape*, pp. 98-188, for the longest discussion to date of how Clare tackled this problem.

45. H. S. Milford, ed., *Cowper: Poetical Works* (London, 1971), p. 201. Hereafter cited as *Cowper: PW*.

46. Cecil Gould, *Space in Landscape* (London, 1974), p. 33.

47. *Countries of the Mind* (London, 1937), p. 71.

Chapter 2. 'Thy wild seclusions'

1. New York, 1970, pp. 10-11. Hereafter cited as *Essay*.

2. *Essays on the Picturesque*, II. 260. Hereafter cited as *Essays*.

3. Ernest de Selincourt, ed., *Letters of William and Dorothy Wordsworth* (Oxford, 1937), I. 244.

4. *The Poetry of Earth* (New York, 1974), p. 42. Although Gray's interest in natural history increased towards the end of his life (after most of his poetry had been written), he was renowned throughout his life for his wide erudition, little of which is incorporated directly into his poetry.

5. Karl Kroeber, ed., *Backgrounds to British Romantic Literature* (San Francisco, 1968), pp. 19-20.

6. Clare did not own a Bewick book, but he knew of his work. In NMS 7, he has scribbled down 'Bewicks Birds' twice as part of a list of desirable books.

7. *Georgic Tradition in English Poetry* (New York, 1935), p. 209.

8. *The Birds of Scotland* (Edinburgh, 1806), pp. 41-2.

9. J. Logie Robertson, ed., *James Thomson: Poetical Works* (London, 1965), p. 29. Hereafter cited as *Thomson: PW*.

10. See *The Idea of Landscape*, pp. 123-4.

11. Geoffrey Keynes, ed., *The Complete Writings of William Blake* (London, 1966), p. 687.

12. From a manuscript at Yale published by A. J. V. Chapple in *The Yale University Library Gazette*, 31. 1 (July 1956), 48.

13. The comparison is, of course, unjust to Wordsworth. Not only is the sonnet taken out of its context as part of 'The River Duddon' sequence, but it is obviously one of Wordsworth's weaker pieces. The comparison should reveal, however, where the true strength of each poet lies.

14. The Tibble version of this sonnet attempts to put these moving pictures into the frames of punctuation, but Clare's vitality remains. Their attempt to divide this poem into clauses is in vain, and the addition of eight commas, two colons, one semicolon, and three full stops is merely irritating. The poem needs punctuation no more than a Bewick needs to be framed; they are both best left surrounded by emptiness. For a woodcut of geese attacking a girl beside a stream with stepping-stones, see *1800 Woodcuts by Thomas Bewick and his School*, ed. Blanche Cirker (New York, 1962), Plate 165, No. 5.

15. *The Crowning Privilege* (London, 1955), p. 51.

16. Wordsworth's poem on the Pleasure-Ground was written in 1814 but not published until 1827. Clare may have known the poem, but 'Walcott Hill and Surrounding Scenery' was probably not indebted to Wordsworth as it was written between 1822 and 1824.

17. *Emblem and Expression* (London, 1975), p. 26.

18. The sentiments expressed in this poem are close to those in an early Byron poem, 'Inscription on the Monument of a Newfoundland Dog'. (*The Works of Lord Byron*, ed. E. H. Coleridge (London, 1903), I. 280-1.) Clare admired Byron and was to write his own *Don Juan* and *Child Harold* in the Asylum.

Chapter 3. 'Adams open gardens'

1. *Lives of the Poets* (London, 1785), VI. 173-4.

2. *Poems by John Clare* (London, 1908), p. 19.

3. London, M. Cooper, 1756, p. 51.

4. Eric Robinson and Geoffrey Summerfield, eds., *John Clare: The Shepherd's Calendar* (London, 1964), pp. 46-7. Hereafter cited as *SC*.

5. Published by Geoffrey Grigson in *The Mint* (1946), p. 175.

6. *The Poetry of John Clare: A Critical Introduction* (London, 1974), pp. 69-84.

7. Allan Wade, ed., *The Letters of W. B. Yeats* (London, 1954), p. 466.

8. Nikolaus Pevsner, *Studies in Art, Architecture and Design*, I. 94.

9. Nikolaus Pevsner, ed., *The Picturesque Garden and its Influence outside the British Isles*, p. 60.

10. Leslie Parris, Ian Fleming-Williams, and Conal Shields, *Constable: Paintings, Watercolours and Drawings* (London, 1976), pp. 171-2.

11. Jack Lindsay, *Turner* (St. Albans, 1973), p. 288.

12. Mary Moorman, ed., *Journals of Dorothy Wordsworth* (London, 1971), p. 9. Cf. Wordsworth's great passage on mist in Book XIII (1805) of *The Prelude*, which is actually a definition of Imagination.

13. *Unprofessional Essays* (London, 1956), p. 106.

14. *Studies in Romanticism*, 14. 3 (Summer 1975), 278, 276.

15. *English Gardens and Landscapes: 1700-1750* (London, 1967), p. 15.

16. *A Memoir of Thomas Bewick: Written by Himself* (London, 1975), p. 27. Bewick died in 1828 but the *Memoir* was not published until 1862. Ruskin praised it highly, and it would have delighted Clare.

17. Marshall McLuhan and Quentin Fiore, *The Medium is the Massage* (New York, 1967), unpaginated.

18. Martin Hardie, *Water Colour Painting in Britain* (London, 1967), II. 51.

19. *The Shepherd's Calendar; with Village Stories, and other Poems* (London, 1827), p. 90. Tibble reprints this verse as in Taylor, but makes changes in other verses. This comparison is intended to stress my opinion that the 1964 version is superior; I do not want to imply that Constable's finished pictures are less interesting or less artistic than his sketches.

20. *Eclogues and Georgics*, trans. T. F. Royds (London, 1965), pp. 81-2.

21. In a letter to the author, January 1978.

22. *Art and Illusion: a Study in the Psychology of Pictorial Representation* (New Jersey, 1972), p. 172.

23. Readers of Jane Austen are quickly made aware of how walks help to develop her characters, and how walks are interwoven with her plots.

24. A. Norman Jeffares, ed., *Cowper: Selected Poems and Letters* (London, 1963), p. 125.

Chapter 4. 'Natures wild Eden'

1. Clare's second visit to London took place during May and June 1822. There is no information in the Royal Academy catalogues about what was on display during those months, as the regular exhibitions were later in the summer. What he actually saw during his London visits is therefore difficult to ascertain. If he never saw a painting by Claude, he was certainly aware of the neo-classical style. His other London visits took place during March and April 1820, May and June 1824, and February and March 1828. See *Prose*, p. 83 and J. W. and Anne Tibble, *John Clare: A Life* (London, 1972), pp. 117-19, 169-91, 201-18, 254-8.

2. Francis Greenacre, *The Bristol School of Artists: Francis Danby and Painting in Bristol 1810-1840* (Bristol, 1973), pp. 9-32, 121-39.

3. See Eric Adams, *Francis Danby: Varieties of Poetic Landscape* (London, 1973). Hereafter cited as *Francis Danby*.

4. J. W. and Anne Tibble, *John Clare: A Life*, p. 171.

5. Either Clare was mistaken or Rippingille was an unusually precocious artist, being 11 years old in 1809.

6. BL MS Eg. 2246, fo. 366.

7. See the catalogue produced by the Usher Gallery, Lincoln, *Peter De Wint, 1784-1849* (St. Ives, no date given); and Martin Hardie, *Water Colour Painting in Britain*, Vol. II (London, 1967).

8. See *Shock of Recognition: the Landscape of English Romanticism and the Dutch seventeenth-century school* (Arts Council of Great Britain, London, 1971).

9. M. Kitson and A. Wedgwood, *English Painting* (London, 1964), p. 14.

10. See Dorothy Stroud, *Capability Brown* (London, 1975), pp. 74-9, 145, 207; plates 12a, b, 13a, b, 14a, b.

11. See *Guide to Burghley House* (Stamford, no date given), available from the Agent, Burghley House, Stamford.

12. John Dixon Hunt and Peter Willis, eds., *The Genius of the Place* (London, 1975), p. 344.

13. *Capability Brown*, pp. 117-18.

14. *The Genius of the Place*, p. 377.

15. *Capability Brown*, p. 202.

16. *The Genius of the Place*, pp. 361, 363.

17. See Paul Fussell, *The Rhetorical World of Augustan Humanism* (Oxford, 1965), Chapter 9.

18. For two useful discussions of this aspect of *Mansfield Park*, see Alistair M. Duckworth, *The Improvement of the Estate* (The Johns Hopkins University Press, 1971), Chapter 1; and Alan Kennedy, *Meaning and Signs in Fiction* (Macmillan, 1979), Chapter 4.

19. R. B. Beckett, ed., *John Constable's Discourses* (Ipswich, 1970), p. 59.

20. *The Art of Claude Lorrain* (London, 1969), p. 17.

21. *Landscape into Art* (Harmondsworth, 1966), p. 46.

22. Andrew H. Wright, *Jane Austen's Novels* (Harmondsworth, 1962), p. 17.

23. Arthur Pollard, ed., *Crabbe: The Critical Heritage* (London, 1972), p. 284.

24. C. P. Barbier, *William Gilpin* (London, 1963), p. 71.

25. Arthur Sherbo, ed., *Memoirs and Anecdotes of Dr. Johnson* (London, 1974), p. 93.

26. Huon Mallalieu, *The Norwich School* (London, 1974), pp. 52, 50, 44, 27, 31, 22, 23, Colour Plate III, 25.

27. The early fortunes of this fine picture are an indication of how little landscape without incident was appreciated. While in the possession of an early owner, it fell into two pieces, which he used as studio blinds. Later, an artist called Bristow 'enlivened' it with a shepherd boy and sheep. Derek and Timothy Clifford, *John Crome* (Greenwich, Connecticut, 1968), pp. 96, 100, Plate 111.

28. Kenneth Clark, *Civilisation* (London, 1970), p. 282.

29. Ian Jack, *Keats and the Mirror of Art* (Oxford, 1967), pp. 21-2.

30. Robin Reilly, *British Watercolours* (London, 1974), p. 47.

Chapter 5. 'Rich instincts natural taste'

1. Jean H. Hagstrum, *The Sister Arts* (Chicago, 1958), p. iii.

2. R. T. Davies and B. G. Beatty, eds., *Literature of the Romantic Period 1750-1850* (Liverpool, 1976), p. 82.

3. John Nolen, ed., *The Art of Landscape Gardening* (London, 1907), p. 67.

4. Mary Moorman, *William Wordsworth: The Later Years 1803-1850* (London, 1968), p. 370.

5. T. H. Green and T. H. Grose, eds., David Hume, *Essays Moral, Political and Literary* (London, 1912), I. 240.

6. Harold Bloom and Lionel Trilling, eds., *Romantic Poetry and Prose* (New York, 1973), p. 597.

7. *The Rhetorical World of Augustan Humanism* (Oxford, 1965), p. 234.

8. Dennis Davison, ed., *The Penguin Book of Eighteenth-Century English Verse* (Harmondsworth, 1973), pp. 236-7.

9. *The Rhetorical World of Augustan Humanism*, p. 240.

10. Conor Cruise O'Brien, ed., *Reflections on the Revolution in France* (Harmondsworth, 1969), pp. 192-3.

11. Herbert Davis, ed., *Swift: Poetical Works* (London, 1967), pp. 588-9.

12. James Gibson, ed., *The Complete Poems of Thomas Hardy* (London, 1981), pp. 146-7.

13. Richard Ellmann and Robert O'Clair, eds., *The Norton Anthology of Modern Poetry* (New York, 1973), p. 445.

14. *The Rhetorical World of Augustan Humanism*, p. 236.

15. Basil Taylor, *Constable: Paintings, drawings and watercolours* (London, 1973), p. 224.

16. Robert Gittings, ed., *Letters of John Keats* (London, 1970), p. 229.

17. *Three Essays: on Picturesque Beauty; on Picturesque Travel; and on Sketching Landscape* (London, 1794), pp. 54, 57, 58.

18. 'The Tulip and the Bee'. PMS, A 51-92. See Eric Robinson, ed., *Clare's Countryside* (London, 1981), p. 59.

SELECT BIBLIOGRAPHY

(Dates of original publication appear in square brackets)

ADAMS, ERIC. *Francis Danby: Varieties of Poetic Landscape.* London: Yale University Press, 1973.

ADLARD, JOHN. 'John Clare: The Long Walk Home'. *English,* Vol. 19 (Autumn 1970), pp. 85-9.

AIKIN, JOHN. *An Essay on the Application of Natural History to Poetry.* New York: Garland Publishing, 1970 [1777].

APPLETON, JAY. *The Experience of Landscape.* London: John Wiley and Sons, 1975.

ARTS COUNCIL OF GREAT BRITAIN. *Shock of Recognition: The Landscape of English Romanticism and the Dutch Seventeenth-Century School.* London: Arts Council of Great Britain, 1971.

AUBIN, R. A. *Topographical Poetry in XVIII-Century England.* New York: The Modern Language Association of America, 1936.

BARRELL, JOHN. *The Idea of Landscape and The Sense of Place 1730-1840: an approach to the Poetry of John Clare.* Cambridge: Cambridge University Press, 1972.

BATE, W. J. *From Classic to Romantic.* Cambridge, Mass.: Harvard University Press, 1946.

BEWICK, THOMAS. *A Memoir of Thomas Bewick: Written by Himself.* London: Oxford University Press, 1975 [1862].

BLOOM, HAROLD. *The Visionary Company: A Reading of English Romantic Poetry.* Ithaca: Cornell University Press, 1971 [1961].

—— and LIONEL TRILLING, eds. *Romantic Poetry and Prose.* London: Oxford University Press, 1973.

BLUNDEN, EDMUND. *Nature in English Literature.* London: Hogarth Press, 1929.

BROWN, R. W. 'John Clare's Library'. *Northamptonshire Natural History Society and Field Club,* Vol. 25, No. 199 (September 1929), pp. 56-64.

BROWNLOW, TIMOTHY. 'A Molehill for Parnassus: John Clare and Prospect Poetry'. *University of Toronto Quarterly,* Vol. 48, No. 1 (Fall 1978), pp. 23-40.

—— Review of *The Poetry of John Clare: a Critical Introduction* by Mark Storey (Macmillan, 1974). *Notes and Queries,* NS Vol. 23, No. 3 (March 1976), pp. 135-7.

BURKE, EDMUND. *A Philosophical Enquiry into the Origin of our Ideas of the Sublime and Beautiful*, ed. J. T. Boulton. London: Routledge and Kegan Paul, 1958 [1757].

BURNEY, E. L. 'Peasant poets, or peasants, poetry, patronage and the pauper's pit'. *The Manchester Review* (Autumn 1966), pp. 59-72.

CAMPBELL, BRUCE. 'The Birds of John Clare'. *Folio* (Summer 1980), pp. 3-5.

CHAPPLE, A. J. V. 'Some unpublished Poetical Manuscripts of John Clare'. *The Yale University Library Gazette*, Vol. 31, No. 1 (July 1956), pp. 34-48.

CHRIST, CAROL T. *The Finer Optic: the Aesthetic of Particularity in Victorian Poetry*. New Haven: Yale University Press, 1975.

CLARE, JOHN. *Bird Poems*, introduction Peter Levi. London: The Folio Society, 1980.

—— *Birds Nest: Poems by John Clare*, ed. James Kirkup and Anne Tibble. Northumberland: The Mid Northumberland Arts Group, 1973.

—— *Clare's Countryside*, ed. Eric Robinson, introduction Brian Patten. London: Heinemann/Quixote Press, 1981.

—— *The Journals, Essays, and the Journey from Essex*, ed. Anne Tibble. Manchester: Carcanet New Press, 1980.

—— *The Letters of John Clare*, ed. J. W. and Anne Tibble. London: Routledge and Kegan Paul, 1970 [1951].

—— *The Midsummer Cushion*, ed. Anne Tibble and R. K. R. Thornton. Northumberland: The Mid Northumberland Arts Group, 1979.

—— *Poems*, ed. Arthur Symons. London: Henry Frowde, 1908

—— *The Poems of John Clare*, 2 vols., ed. J. W. Tibble. London: J. M. Dent, 1935.

—— *Poems Chiefly from Manuscript*, ed. Edmund Blunden and Alan Porter. London: Richard Cobden-Sanderson, 1920.

—— *Poems Descriptive of Rural Life and Scenery*. London: Taylor and Hessey, 1820.

—— *The Prose of John Clare*, ed. J. W. and Anne Tibble. London: Routledge and Kegan Paul, 1970 [1951].

—— *The Rural Muse*. London: Whittaker, 1835.

—— *Selected Poems*, ed. Elaine Feinstein. London: University Tutorial Press, 1968.

—— *Selected Poems of John Clare*, ed. James Reeves. London: Heinemann, 1968 [1954].

—— *Selected Poems and Prose of John Clare*, ed. Eric Robinson and Geoffrey Summerfield. London: Oxford University Press, 1967.

—— *The Shepherd's Calendar*, ed. Eric Robinson and Geoffrey Summerfield. London: Oxford University Press, 1973 [1964].

—— *The Shepherd's Calendar; with Village Stories and Other Poems.* London: John Taylor, 1827.

—— *The Village Minstrel and Other Poems.* London: Taylor and Hessey, 1821.

CLARK, KENNETH. *Landscape into Art.* Harmondsworth: Penguin Books, 1966 [1949].

CLARKE, PHILIP, BRIAN JACKMAN and DERRIK MERCER, eds. *The Sunday Times Book of the Countryside.* London: Macdonald and Jane, 1980.

CLIFFORD, DEREK and TIMOTHY. *John Crome.* Greenwich, Connecticut: New York Graphic Society, 1968.

COBBETT, WILLIAM. *Rural Rides.* Harmondsworth: Penguin Books, 1967 [1830].

COLERIDGE, S. T. *Poetical Works*, ed. E. H. Coleridge. London: Oxford University Press, 1969 [1912].

COMBE, WILLIAM. *The Tour of Doctor Syntax in Search of the Picturesque.* London: Methuen, 1903 [1809].

COWPER, WILLIAM. *Poetical Works*, ed. H. S. Milford. London: Oxford University Press, 1971 [1905].

CROSSAN, G. D. 'John Clare: A Chronological Bibliography'. *Bulletin of Bibliography and Magazine Books*, Vol. 32, No. 2 (April—June 1975), pp. 55-62, 88.

DAY LEWIS, CECIL. *The Lyric Impulse.* Cambridge, Mass.: Harvard University Press, 1965.

DENDURENT, H. O. *John Clare: a reference guide.* Boston: G. K. Hall, 1978.

DRABBLE, MARGARET. *A Writer's Britain: Landscape in Literature.* London: Thames and Hudson, 1979.

DRUCE, G. C. 'Northamptonshire Botanologia: John Clare'. *Northamptonshire Natural History Society and Field Club*, Vol. 16, No. 130 (June 1912), pp. 183-214.

DUCK, STEPHEN. *Poems on Several Occasions.* London: Printed for the Author, 1736.

DURLING, D. L. *Georgic Tradition in English Poetry.* New York: Columbia University Press, 1935.

FISHER, JAMES. 'The Birds of John Clare'. *The First Fifty Years: a History of the Kettering and District Naturalists' Society and Field Club, 1905-55*, pp. 26-69.

—— 'John Clare: Naturalist and Poet'. *The Listener* (19 October 1961), pp. 614-15.

148 BIBLIOGRAPHY

FOSTER, JOHN WILSON. 'A Redefinition of Topographical Poetry'. *Journal of English and Germanic Philology*, Vol. 69 (1970), pp. 394-406.

—— 'The Topographical Tradition in Anglo-Irish Poetry'. *Irish University Review*, Vol. 4, No. 2 (1974), pp. 169-87.

—— 'The Measure of Paradise: Topography in Eighteenth-Century Poetry'. *Eighteenth-Century Studies*, Vol. 9, No. 2 (Winter 1975-6), pp. 232-56.

FROSCH, THOMAS R. 'The Descriptive Style of John Clare'. *Studies in Romanticism*, Vol. 10 (1971), pp. 137-48.

FUSSELL, PAUL. *The Rhetorical World of Augustan Humanism: Ethics and Imagery from Swift to Burke.* Oxford: Clarendon Press, 1965.

GILPIN, ARTHUR. 'John Clare: The Birdwatcher's Poet'. *Country Life* (13 September 1979), pp. 766-7.

GILPIN, WILLIAM. *Observations on the River Wye.* London: R. Blamire, 1782.

—— *Three Essays: on Picturesque Beauty; on Picturesque Travel; and on Sketching Landscape: to which is added a poem, on Landscape Painting.* London: R. Blamire, 1792.

GOMBRICH, E. H. *Art and Illusion: A Study in the Psychology of Pictorial Representation.* New Jersey: University of Princeton Press, 1972 [1960].

GOULD, CECIL. *Space in Landscape.* Themes and Painters in the National Gallery, No. IX. London: Publications Department of the National Gallery, 1974.

GRAHAME, JAMES. *Poems*, 2 vols. London: Longman, Hurst, Rees, and Orme, 1807.

—— *The Birds of Scotland.* Edinburgh: Longman, Hurst, Rees, and Orme, 1806.

GRAINGER, MARGARET. *A Descriptive Catalogue of the John Clare Collection in the Peterborough Museum and Art Gallery.* Peterborough: G. H. Fisher, 1973.

GRAVES, ROBERT. *The Common Asphodel.* London: Hamish Hamilton, 1949.

GREENACRE, FRANCIS. *The Bristol School of Artists: Francis Danby and Painting in Bristol 1810-1840.* Bristol: City Art Gallery, 1973.

GREGOR, IAN. 'The Last Augustan: Some Observations on the Poetry of George Crabbe (1755-1832)'. *The Dublin Review*, Series 4, Vol. 229, No. 467 (First Quarter 1955), pp. 37-50.

GRIGSON, GEOFFREY. 'Integrity and John Clare'. *The Listener* (3 November 1955), pp. 743-4.

—— ed. 'John Clare: Poems and Fragments'. *The Mint* (1946), pp. 170-8.

HAGSTRUM, JEAN H. *The Sister Arts: the Tradition of Literary Pictorialism and English Poetry from Dryden to Gray.* Chicago: The University of Chicago Press, 1958.

HARDIE, MARTIN. *Water Colour Painting in Britain,* 2 vols. London: B. T. Batsford, 1967.

HATLEY, VICTOR A. 'The Poet and the Railway Surveyors, an Incident in the Life of John Clare'. *Northamptonshire Past and Present,* Vol. 5, No. 2 (1974), pp. 101-6.

HEATH-STUBBS, JOHN. 'John Clare and the Peasant Tradition'. *The Penguin New Writing.* London: Penguin Books (1947), pp. 112-24.

HEFFERNAN, J. A. W. 'Wordsworth on the Picturesque'. *English Studies,* Vol. 49 (1968), pp. 489-98.

HERRMANN, LUKE. *British Landscape Painting of the 18th Century.* London: Faber and Faber, 1973.

HIPPLE, W. J. *The Beautiful, The Sublime, and The Picturesque in Eighteenth-Century British Aesthetic Theory.* Carbondale: Illinois University Press, 1957.

HOSKINS, W. G. *The Making of the English Landscape.* Harmondsworth: Penguin Books, 1970 [1955].

HOWARD, WILLIAM. *John Clare.* Boston: Twayne Publishers, 1981.

HUMPHREYS, A. R. *William Shenstone: an Eighteenth-Century Portrait.* Cambridge: Cambridge University Press, 1937.

HUNT, JOHN DIXON. *The Figure in the Landscape: Poetry, Painting, and Gardening during the Eighteenth Century.* Baltimore: The Johns Hopkins University Press, 1976.

—— and PETER WILLIS, eds. *The Genius of the Place: The English Landscape Garden 1620-1820.* London: Elek, 1975.

HUSSEY, CHRISTOPHER. *The Picturesque: Studies in a Point of View.* London: Frank Cass, 1967 [1927].

HYAMS, EDWARD. *The English Garden.* London: Thames and Hudson, 1966.

JACK, IAN. *Keats and the Mirror of Art.* Oxford: Clarendon Press, 1967.

—— 'Poems of John Clare's Sanity'. In *Some British Romantics,* ed. John Jordan, James Logan, Northrop Frye. Dayton: Ohio State University Press (1966), pp. 189-232.

JOHNSON, E. D. H., ed. *The Poetry of Earth: a Collection of English Nature Writings.* New York: Atheneum, 1974 [1966].

KEITH, W. J. *The Poetry of Nature: Rural Perspectives in Poetry from Wordsworth to the Present.* Toronto: University of Toronto Press, 1980.

—— *The Rural Tradition: William Cobbett, Gilbert White and other non-fiction prose writers of the English countryside.* Toronto: University of Toronto Press, 1975.

KNIGHT, RICHARD PAYNE. *The Landscape, a Didactic Poem.* London: W. Bulmer, 1794.

KROEBER, KARL, ed. *Backgrounds to British Romantic Literature.* San Francisco: Chandler Publishing Co., 1968.

LINDSAY, JACK. *Turner.* St. Albans: Panther Books, 1973.

LISTER, RAYMOND. *British Romantic Art.* London: G. Bell, 1973.

LUPINI, BARBARA. 'An Open and Simple Eye: The Influence of Landscape in the Work of John Clare and Vincent Van Gogh'. *English*, Vol. 23 (Summer 1974), pp. 58-62.

McLUHAN, MARSHALL and HARLEY PARKER. *Through the Vanishing Point.* New York: Harper and Row, 1968.

MALINS, EDWARD. *English Landscaping and Literature 1660-1840.* London: Oxford University Press, 1966.

MALLALIEU, HUON. *The Norwich School: Crome, Cotman and their Followers.* London: Academy Editions, 1974.

MANWARING, E. W. *Italian Landscape in Eighteenth Century England.* London: Frank Cass, 1965 [1925].

MURRY, J. MIDDLETON. *The Problem of Style.* London: Oxford University Press, 1960 [1922].

NABHOLTZ, J. R. 'Dorothy Wordsworth and the Picturesque'. *Studies in Romanticism*, Vol. 3 (1964), pp. 118-28.

—— 'The *Guide to the Lakes* and the Picturesque Tradition'. *Modern Philology*, Vol. 61, No. 4 (May 1964), pp. 288-97.

NOYES, RUSSELL. *Wordsworth and the Art of Landscape.* Bloomington: Indiana University Press, 1968.

PARRIS, LESLIE, ed. *Landscape in Britain c.1750-1850.* London: Tate Gallery Publications, 1973.

——, IAN FLEMING-WILLIAMS, CONAL SHIELDS. *Constable: Paintings, Watercolours and Drawings.* London: Tate Gallery, 1976.

PEAKE, CHARLES, ed. *Poetry of the Landscape and the Night: Two Eighteenth-Century Traditions.* London: Edward Arnold, 1967.

PEVSNER, NIKOLAUS. *Studies in Art, Architecture and Design,* Vol. 1: *From Mannerism to Romanticism.* London: Thames and Hudson, 1968.

—— ed. *The Picturesque Garden and its Influence outside the British Isles.* Washington: Dumbarton Oaks, 1974.

POPE, ALEXANDER. *The Poems of Alexander Pope*, ed. John Butt. XI vols. London and New Haven, 1939-61.

POWELL, DAVID. *Catalogue of the John Clare Collection in the Northampton Public Library.* Northampton: John Dickens, 1964.

PRAZ, MARIO. *The Romantic Agony.* London: Oxford University Press, 1970 [1933].

PRICE, MARTIN, ed. *The Restoration and the Eighteenth Century.* London: Oxford University Press, 1973.

—— 'The Picturesque Moment'. In *From Sensibility to Romanticism*, ed. F. W. Hilles and Harold Bloom. New York: Oxford University Press (1965), pp. 259-92.

PRICE, UVEDALE. *Essays on the Picturesque.* London: J. Mawman, 1810.

PRINCE, HUGH. *Parks in England.* Isle of Wight: Pinhorns, 1967

QUENNELL, PETER. *Romantic England: Writing and Painting 1717-1851.* London: Weidenfeld and Nicolson, 1970.

REPTON, HUMPHRY. *Fragments on the Theory and Practice of Landscape Gardening.* London: J. Taylor, 1816.

REYNOLDS, GRAHAM. *A Concise History of Watercolours.* London: Thames and Hudson, 1971.

—— *Constable: the Natural Painter.* St Albans: Panther, 1976 [1965].

REYNOLDS, JOSHUA. *Discourses on Art*, ed. R. R. Wark. London: Collier-Macmillan, 1969 [1797].

ROBINSON, ERIC and GEOFFREY SUMMERFIELD. 'John Clare (1793-1864): A Poet in his Joy'. *The Cambridge Review*, Vol. 86 (23 January 1965), pp. 194-9.

—— 'John Taylor's editing of Clare's *The Shepherd's Calendar*'. *Review of English Studies*, NS Vol. 14, No. 56 (November 1963), pp. 359-69.

—— 'Unpublished poems by John Clare'. *The Listener* (29 March 1962), pp. 556-7.

—— 'Unpublished poems by John Clare'. *The Malahat Review*, No. 2 (April 1967), pp. 106-20.

STOREY, MARK, ed. *Clare: The Critical Heritage.* London: Routledge and Kegan Paul, 1973.

—— *The Poetry of John Clare: A Critical Introduction.* London: Macmillan, 1974.

—— Review of *In Adam's Garden: a Study of John Clare's Pre-Asylum Poetry* by Janet M. Todd. University of Florida Press, 1973. *Essays in Criticism*, Vol. 24, No. 4, pp. 399-406.

STROUD, DOROTHY. *Capability Brown.* London: Faber and Faber, 1975 [1950].

SUTHERLAND, JAMES. *A Preface to Eighteenth Century Poetry.* London: Oxford University Press, 1966 [1948].

SWINGLE, L. J. 'Stalking the Essential John Clare: Clare in Relation to His Romantic Contemporaries'. *Studies in Romanticism*, Vol. 14, No. 3 (Summer 1975), pp. 273-84.

THOMSON, JAMES. *Poetical Works*, ed. J. Logie Robertson. London: Oxford University Press, 1965 [1908].

TIBBLE, J. W. and ANNE. *John Clare: A Life.* London: Michael Joseph, 1972 [1932].

TINKER, C. B. *Nature's Simple Plan: A Phase of Radical Thought in the Mid-Eighteenth Century.* New Jersey: Princeton University Press, 1922.

TODD, JANET M. *In Adam's Garden: A Study of John Clare's Pre-Asylum Poetry.* University of Florida Humanities Monograph No. 39. Gainesville: University of Florida Press, 1973.

—— 'John Clare: a Bibliographical Essay'. *The British Studies Monitor*, Vol. 4, No. 2, Issue II (Winter 1974), pp. 3-18.

UNWIN, RAYNER. *The Rural Muse: Studies in the Peasant Poetry of England.* London: George Allen and Unwin, 1954.

VIRGIL. *Eclogues and Georgics*, trans. T. F. Royds. London: Dent, 1965.

WARNER, ALAN. 'Stephen Duck, the Thresher Poet'. *A Review of English Literature*, Vol. 8, No. 2 (April 1967), pp. 38-48.

WARTON, JOSEPH. *An Essay on the Writings and Genius of Pope.* London: M. Cooper, 1756.

WATSON, J. R. *Picturesque Landscape and English Romantic Poetry.* London: Hutchinson Educational, 1970.

WHITE, GILBERT. *The Journals of Gilbert White*, ed. Walter Johnson. New York: Taplinger Publishing, 1970 [1931].

—— *The Natural History of Selborne.* London: Oxford University Press, 1971 [1788].

WILLIAMS, RAYMOND. *The Country and the City.* London: Chatto and Windus, 1973.

WORDSWORTH, DOROTHY. *Journals of Dorothy Wordsworth*, ed. Mary Moorman. London: Oxford University Press, 1971.

WORDSWORTH, WILLIAM. *Guide to the Lakes*, ed. Ernest de Selincourt. London: Oxford University Press, 1970 [1810].

—— *The Poetical Works of William Wordsworth*, ed. Ernest de Selincourt. 5 vols. Oxford: Clarendon Press, 1940-9.

—— *The Prelude*, ed. Ernest de Selincourt. London: Oxford University Press, 1959 [1850].

WRIGHT, DAVID, ed. *The Penguin Book of English Romantic Verse.* Harmondsworth: Penguin Books, 1970.

GLOSSARY

awe: variant of *haw*
baulk: a narrow strip of grass dividing two ploughed fields
bent: coarse grass
brig: variant of *bridge*
brunt: to make an abrupt or forceful movement
clipping pinks: flowers presented to sheep-shearers
clown: a rustic
cotter: variant of *cottager*
cowslap peeps: cowslip eyes
crick: to bounce under the leg?
croodling: contracting the body from cold; shrinking or huddling
curdle up: to rise in rings, move in a circular fashion
do: ditto
dotterel: a pollard tree
drabbled: dirtied or splashed with walking in the mud
eke: to add to, to increase
elting: soft, ploughed
flirt: to flit or flutter; n. fluttering
gelid: freezing cold
hirkle: to crouch, to set up the back, as cattle who shrink from the cold
hugh: Clare's spelling of *huge*
jilt: to throw underhand with a quick and suddenly arrested motion
lads love: southernwood
mawl: to drag along wearily
mozzling: mottled
neer: nearly
padded: well trodden
peeps: see *cowslap peeps*
pettichap: the chiffchaff or willow-warbler
pingle: a close, small meadow
pismire: the ant
platt or *plat:* a flat stretch of ground
pooty: landsnail, *Capaea*, particularly the shell of *C. nemoralis*
prevade: variant of *pervade*
prog: to poke or prod
puddock: kite or buzzard
pudgy: full of puddles

ramp: to grow luxuriantly
rawky: misty, foggy
reak: to emit steam or vapour
sallow: species of *Salix*
sinkfoil: cinquefoil
skriecker: a wooden rattle
spindle: the shoot or stem of a plant
snuft: to snuffle or sniff
starnel: starling
struttle: stickleback
stulp: stump
sturt: to move suddenly
swail: shade
swathy: striped?
sweeing or *at swee:* swaying, rocking
taws: marbles
toze: to pluck, to snatch
twitch: spear grass, couch-grass
waterblob: marsh marigold
whimpling: rippling, meandering

INDEX

158   INDEX

Shakespeare, William 42, 43
Shelley, P. B. 6, 87
Shenstone, William 15, 120
Simonides 116
Smart, Christopher 3
Smith, Bernard 45, 46
Sowerby, James 24
Spenser, Edmund 69
Stendhal 5, 14
Storey, Mark 75, 137
Stourhead, Wilts. 15, 65, 66
Stowe, Bucks. 15, 65, 66, 81
Stubbs, George 97
Sublime, the 25, 104, 135
Summerfield, Geoffrey 7, 39, 92, 93, 137, 138, 140
Swift, Jonathan 26, 121, 143
Swingle, L. J. 87
Symons, Arthur 68

Taste 3, 6, 12, 37-9, 44, 47, 51, 53-4, 61, 64, 72, 93, 103, 116-33
Tate Gallery, The 97
Tatersal, Robert 27
Taylor, John 2, 23, 39, 92, 140
Temple, Sir William 15
Thackeray, W. M. 108
Theocritus 4, 98
Thirtle, John 112
Thomson, James 3, 5, 25, 27, 35, 44, 51, 69, 70, 72-4, 82, 83, 95, 118, 139
Thoreau, H. D. 26
Thornton, John 24
Thornton, R. K. R. 135, 136
Thorpe, W. H. 127, 128

Tibble, Anne 135, 136, 141
Tibble, J. W. 7-8, 22, 135-7, 139, 140, 141
Turner, J. M. W. 5, 83, 93, 98, 140

ut pictura poesis 97, 116

Vanbrugh, Sir John 47
Vaughan, Henry 67
Versailles 87
Virgil 3, 4, 44, 94

Walpole, Horace 18, 81, 97
Walton, Izaak 100
Warton, Joseph 69, 70, 82
Warton, Thomas 31
Water-colour painting 24, 97, 110-12, 116
Waterloo, Antonie 109
Watson, J. R. 6, 18
West, Benjamin 92
White, Gilbert 6, 26, 29, 45, 48, 49, 53-6, 85, 125
Wilkie, David 71
Wilton House, Wilts. 97
Woburn, Beds. 97
Woodhouse, James 27
Wordsworth, Dorothy 63, 83, 138, 140
Wordsworth, William 1, 3, 4, 6, 10, 11, 14, 16-19, 25, 27, 39, 44, 57-9, 62-4, 66, 72, 80, 82, 83, 86, 91, 95, 104, 113, 114, 117-19, 122, 125, 127, 136-40, 142
Wotton, Sir Henry 22

Yearsley, Ann 27
Yeats, W. B. 11, 79, 80, 140